OUR WORLD

Martin Newton

NELSON
CENGAGE Learning

Australia • Brazil • Japan • Korea • Mexico • Singapore • Spain • United Kingdom • United States

Our World: Geography concepts and case studies for New Zealand students
1st Edition
Martin Newton

Cover design: Cheryl Rowe, Macarn Design
Text designer: Cheryl Rowe, Macarn Design
Production controller: Siew Han Ong
Reprint: Natalie Orr

Any URLs contained in this publication were checked for currency during the production process. Note, however, that the publisher cannot vouch for the ongoing currency of URLs.

Acknowledgements
The author and publishers wish to thank the following:
NOAA for diagram on page 8; US Navy for images on pages 9 and 10; Geo Eye for image on page 12/13; LINZ for map on page 85; NIWA for maps on page 18; UNDP for image on page 35 (right); Ministry of Economic Development for map on page 98; NASA for satellite images on pages 86, 89, 90 and 146; Alexander Turnbull Library for the cartoon on page 98; Greenpeace NZ for image on page 99; Air New Zealand for image on page 114 (bottom); Cliff Whiting and Metservice (Meteorological Service of New Zealand Limited) for mural on page 116; Tom Scott for cartoon on page 157; Shutterstock for images on pages 7, 8 (top), 14, 16 (all), 17, 21, 22 (both), 23, 24 (bottom), 25 (all), 27 (both), 28 (both), 29 (all), 32/33, 34 (all), 35 (left and bottom), 36, 38, 39 (top and middle), 42, 44, 47, 48,49,50, 52 (both), 53 (all), 54, 56 (both), 57, (both), 58 (all), 59 (both), 60/61, 62, 63, 64 (all), 67 (top, bottom right), 68 (bottom), 73 (all), 74 (both), 75 (all), 76 (all), 80 (both), 81 (both), 84 (top), 86 (bottom), 91 (all), 92/93, 94, 95 (both), 96 (both), 97 (both), 98 (top right), 101 (both), 102, 103 (both), 104, 105, 106/107, 108 (both), 109 (all), 110 (both), 111, 112 (all), 113 (both), 114 (top), 115, 117, 120 (bottom), 121 (all), 123 (all), 124,126/7, 128, 130, 131, 132 (both) 133, 134, 135, 136,137,138/9, 140, 141, 142,143, 144 (bottom right), 148, 149 (left and bottom), 150 (top left, middle right, 154 156 (all), 163, 164 and 165.

For product information and technology assistance,
in Australia call 1300 790 853;
in New Zealand call 0800 449 725

For permission to use material from this text or product, please email aust.permissions@cengage.com

National Library of New Zealand Cataloguing-in-Publication Data
Newton, Martin.
Our world : geography concepts and case studies for NZ students / Martin Newton.
ISBN 978-017021-570-1
1. Geography. 2. Geography—Case studies. I. Title.
910.76—dc 22

Cengage Learning Australia
Level 7, 80 Dorcas Street
South Melbourne, Victoria Australia 3205

Cengage Learning New Zealand
Unit 4B Rosedale Office Park
331 Rosedale Road, Albany, North Shore 0632, NZ

For learning solutions, visit cengage.co.nz

Printed in Australia by Ligare Pty Limited.
4 5 6 7 8 9 10 20 19 18 17 16

Contents

Geography is all about the living, breathing essence of the world we live in. It explores the past, illuminates the present and prepares us for the future. What could be more important than that?

Michael Palin (president of the Royal Geographical Society, travel writer, television presenter and actor)

Table of concepts

Table of case studies

Overview

We live in a world of geography. Geography is around us all the time. News headlines may not say 'geography' but geography lies behind topics like global warming and sustainability, urban design and ways of providing the food and water needed for the growing world population. Geography aims to describe and explain the world we live in, and also help provide solutions to global problems and help shape a better future world.

Our world is constantly changing. Some changes are slow. Mountains get pushed up only to be eroded down again. Eroded material gets carried by rivers to the coast and deposited into the sea to start the slow process of future rock formation. Earthquakes, volcanic activity and landslides can bring about sudden and violent landscape change. People are affected by all these changes. Some we can adapt to, others are beyond our ability to respond to other than to look in awe at the power of nature.

We use the resources of the planet. We farm and mine to provide the food we need to live and the factories the raw materials they need to produce the goods we consume and use. People travel for business and tourism and help establish global links through trade. We create new landscapes through our use of technology to dam rivers, irrigate farmland and construct transport networks. The Internet and other electronic communications shrink the barrier of distance and bring the world into our homes through the TV and computer. Change is so rapid that it can be difficult to keep pace. Facts learnt today are often out of date by tomorrow. Our world is dynamic and we are part of this dynamic world of geography.

This book has a focus on:
1 Concepts that are of geographic significance
2 Case studies from a range of locations and at a range of scales from New Zealand and around the world.

Concepts are big ideas and general understandings that are of lasting value and do not change as rapidly as the facts and figures of our world. Concepts can be thought of as coathangers on which knowledge and facts can be placed. Taking the concept of 'environment' as an example: environments are constantly changing in their make-up and nature, but understanding what environments consist of and how the parts connect together helps us to understand our world. Concepts provide a big picture way of looking at our world.

The book chapters are built around concepts. Each chapter begins with a general overview of the concept, including a definition. This overview is followed by three case studies that illustrate the concept in action. The case studies highlight connections between the natural and cultural worlds. Geographers may make studies of 'physical' and of 'human' geography but in reality the world reflects the connections between people and the natural environment: they are intertwined and depend very much one on the other.

The first and last chapters use a study of real world disaster (Apocalypse) and a fictional story (Hidden) based on fact, to show how within a single event many linkages and concepts are involved.

The book is designed for multi-level use in geography in Years 11, 12 and 13. When understanding or review of a particular concept is required then the relevant chapter of the book could be used at any level. The case studies could be integrated into the geography teaching and learning programme of the school: the first case study from each concept could, for example, be used in Year 11 studies; the second used in Year 12 and the third used in Year 13. The flexibility remains, however, to use case studies in any order and match them to the themes and topics being taught within the school at particular year levels.

ISBN: 9780170215701

Apocalypse: Japan in the news

Sometimes events are so large that they change the geography of places. The huge earthquake and tsunami that struck Japan in 2011 was such an event. This event made news headlines around the world over many days, and provided spectacular images of the power of nature and the devastation nature can bring to places, people and communities. Physical geography and cultural geography were both very much part of this event. To fully understand the event as it played out requires understanding of many important concepts (ideas) of geography.

This event is the focus of Chapter 1. Reports and information about the event are presented in a jumbled way. You need to think of yourself as a news reporter who has received this information with the task of presenting the information for the public to understand.

The Event:

The area is quiet. The absence of people is an eerie thing. Many have died; a few bodies still remain, some were swept out to sea. Those who escaped injury are now living in evacuation centres, and don't know if or when they will be able to return to their villages and homes. Their main concern is neither the cause or size of the disaster but for the well-being and safety of family and friends. Some have lost everything they ever owned and every person they ever loved.

Japan was more prepared for earthquakes and tsunamis than any other country on the planet, yet tens of thousands of people died, buildings collapsed and the landscape of the northern coastline is now completely unrecognisable.

ISBN: 9780170215701

The death toll continues to grow. As many as 20 000 are thought to have died along a coastal strip 200 km long and 10 km wide. Travelling from a landscape untouched by the earthquake and tsunami into the disaster area is like stepping into a ruined world. A local taxi driver described entering the coastal zone as 'like moving from heaven into hell'. Dark brown mud is everywhere, piled high across roads or filling the ground floor of houses, now abandoned. Roofs lie scattered across the ground where buildings have collapsed. Cars can be found high up trees or inside buildings where the tsunami wave dumped them.

One disaster followed the other. The huge earthquake was centred 130 km to the northeast of Japan, causing immediate damage to roads and buildings. Minutes later a ten metre high wall of water – triggered by the undersea earth movements – ploughed towards the coast and then swept inland, destroying everything in its path. The quake and tsunami dealt a fatal blow to the Fukushima nuclear power plant, and radioactive contamination was added to the deadly mix. Nuclear meltdown threatened to strike anyone living in the vicinity of the power plant, and an area of 20 km around the plant was evacuated.

At magnitude 9, Japan's 2011 earthquake is one of the top ten largest ever recorded. It is known by several names: the Sendai earthquake; the Tohoku earthquake; or the Great East Japan Earthquake. Smaller aftershocks continued for days and months after the first main shake.

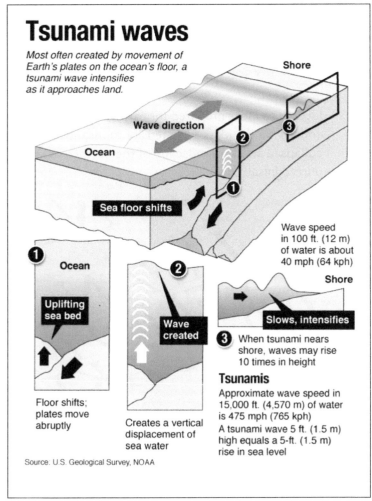

Tsunami waves

Most often created by movement of Earth's plates on the ocean's floor, a tsunami wave intensifies as it approaches land.

Shore

Wave direction

Ocean

Sea floor shifts

2

3

1

Wave speed in 100 ft. (12 m) of water is about 40 mph (64 kph)

Shore

1 Ocean

Uplifting sea bed

Floor shifts; plates move abruptly

2 Wave created

Creates a vertical displacement of sea water

Slows, intensifies

3 When tsunami nears shore, waves may rise 10 times in height

Tsunamis

Approximate wave speed in 15,000 ft. (4,570 m) of water is 475 mph (765 kph)
A tsunami wave 5 ft. (1.5 m) high equals a 5-ft. (1.5 m) rise in sea level

Source: U.S. Geological Survey, NOAA

ISBN: 9780170215701

A man came running across the car park, shouting 'It's coming!', but by then it was too late for people to do anything. The only thing they could do was to go to their upper floors or climb onto roofs, and wait,
hope and pray. Some watched in horror as buildings, cars and rubble were swept inland past their rooftop shelters. Anyone caught in the wave had no chance of survival. Live pictures captured the moment that the wall of water broke over coastal defenses and headed inland.

The earth movement along the fault where the Pacific and North American tectonic plates meet under Japan unleashed power 8000 times more violent than the 2010 and 2011 Christchurch earthquakes. The upward lifting of the seabed where the quake took place first pushed up a huge column of seawater. This body of water then collapsed into itself, generating a great wave that spread quickly away from the quake centre at speeds of up to 800 km/hour. Although tsunami warning sirens sounded along Japan's coast, many people were unable to reach high ground before the wall of water hit 15 minutes later.

Aid was rushed to the devastated area from around Japan and other countries. Long queues began to form for food, medical supplies and petrol. Supplies quickly ran out. No one knew how long it would take for fresh supplies to arrive.

A large ferry boat rests inland amidst destroyed houses in Minatomachi near Miyako city on the east coast of Honshu.

Japan and New Zealand are both located on the edge of the Pacific Ocean on the Pacific Ring of Fire, where giant plate boundaries meet. This is a dangerous place. Around 75% of the world's volcanoes are found around here, and 80% of major world earthquakes are linked to activity in the ring.

ISBN: 9780170215701

A day earlier a visitor to the town would have seen a beautiful bay, fishing port, man-made beaches, sports fields and tourist attractions. Today Minamisanriku, a small city on the east coast of the island of Honshu and 88 km from the earthquake centre, is buried under a sea of mud. The earthquake struck at 2.46 pm local time and 15 minutes later ten metre high waves hit the town. It took just three or four minutes for the tsunami to destroy 95% of the town. Only the tallest buildings remain. An estimated 9500 people (half the city population) are missing presumed drowned. Even those who sheltered on their roofs were swept away.

An aerial view of flood and earthquake damage in the Tohoku region with black smoke coming from the Nippon Oil Sendai refinery.

1 Take the role of a reporter working for a newspaper or TV news programme. The news reports and images about the 2011 earthquake and tsunami that struck Japan have come into your newsroom in a jumbled way under the title of 'Apocalypse'. Study the information and write an illustrated article for publication, or a script (including images) for the 6 pm TV news about the event. The report should be between 200–250 words in length or take approximately 60 seconds to read.

2 The photograph on page 11 shows the tsunami wave breaking over the coast at Miyako city, Iwate Prefecture and sweeping inland. Draw three large frames on a full page, labelled A, B and C. In the centre frame (B), draw an annotated (labelled) sketch of the photograph. In the left frame (A), describe or draw what happened out at sea and on Japan's coastal seabed that caused the tsunami. In the right frame (C), describe what happened next.

ISBN: 9780170215701

3 Events that took place in this great earthquake illustrate many geographic concepts. The three columns in the following table – what happened, a description of the event and the key concept it illustrates – have been jumbled. Reorder the information to correctly match each column.

What happened — events	Elaboration	Concept
Local people caught up in the event had feelings that focused on personal tragedy rather than on the actual magnitude of the earthquake and size of the tsunami waves.	Plate movement caused earthquakes which led to the tsunami. The sequence of events once begun was unstoppable. These two natural events were both hugely destructive for people and property in the coastal area.	Connections
Buildings collapsed despite being strongly built and the landscape was made unrecognisable.	TV news cameras captured the onrushing tsunami and broadcast live pictures. Aid was rushed in from across Japan and from the rest of the world.	Change
Earth movements along the meeting point of two plates triggered a series of disasters. The offshore natural forces caused huge destruction onshore.	The tsunami waves grew larger as they entered shallow water near the coast, while the size of the earthquakes and aftershocks got smaller after the first huge jolt.	Perspective
News of the disaster was almost instantaneous around the world. Help arrived quickly from many sources.	Dark brown mud was everywhere, cars dumped up trees and buildings swept off their foundations. Everything in the path of the tsunami was destroyed.	Processes
The earthquakes and tsunami waves both had a pattern to them that was typical of that type of event.	A local taxi driver called the coastal area as being like 'hell' while the main concern of residents was for the safety of family and friends.	Patterns

ISBN: 9780170215701

ISBN: 9780170215701

Change

The Burj Khalifa building, located in the centre of downtown Dubai, casts a long shadow across the city. Opened in 2010 and at 828 metres high it is the tallest manmade structure in the world (Sky Tower in Auckland is 328 metres high at the top of the mast).

ISBN: 9780170215701

GEOGRAPHY DICTIONARY

Change means to make different or alter. In geography change involves any alteration to the natural or cultural environment. Change can be spatial and/or temporal. Change is a normal process in both natural and cultural environments. It occurs at varying rates, at different times and in different places. Some changes are predictable, recurrent or cyclic, while others are unpredictable or erratic. Change can bring about further change.

Looking at the world from above shows how the global landscape changes from place to place. At high altitude natural changes appear most obvious – from land to sea, from lowlands to mountains and from hot desert to cold ice cap. Closer to the earth the influence of people becomes more obvious as farmland, cities, roads and railways become visible.

The landscape itself is also the result of change and is constantly changing. Mountains are pushed slowly up millimetre by millimetre by tectonic forces, only to be worn down again by rivers and ice. The eroded soil and rock gets redistributed onto lowlands or out into the sea. Earthquakes and volcanic eruptions are natural forces that can destroy and build up the landscape in seconds.

People are travelling across the planet constantly. Some are moving long distances as tourists or migrants, others are on a daily commute to work, school or shops. On farms land is being ploughed, crops planted or crops harvested. The cycle of natural and cultural change is continuous across the planet.

ISBN: 9780170215701

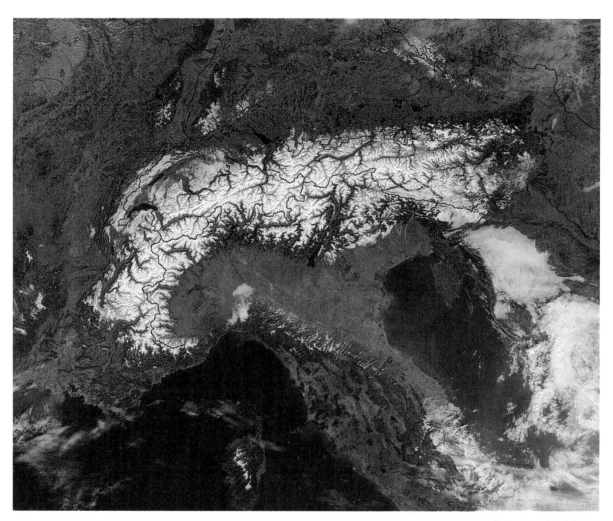

Fig. 2.1 World of change

This satellite photo is a winter image centred on the north of Italy and the snow covered European Alps. The Alps form a border between five nations: Italy, France, Switzerland, Austria and Slovenia. To the west of Italy is the French Mediterranean island of Corsica. The island has snow on its highest mountains as do the highest parts of the Apennine Mountains in Italy. France takes up most of the western side of the image. Germany is to the north. To the east of Italy is the Adriatic Sea. Land on the eastern side of the Adriatic is cloud covered. Although some of the world's great cities – Marseilles (France), Rome, Milan, Venice (Italy) and Geneva (Switzerland) – are within the image area they are less visible than the natural lowland of the North Italian Plain and the dark coloured lakes like Lake Garda, Lake Geneva and Lake Constance.

ISBN: 9780170215701

ISBN: 9780170215701

ACTIVITY

1 a A world of change. Draw a jigsaw map of the world using one of these two jigsaw styles. On each piece of the jigsaw write examples of the way the world changes. Changes from place to place (spatial change) or changes over time (temporal change) could both be included. Use your own ideas and those from Figure 2.1 and the text to help you. Two examples would be: 'LAND TO SEA' (spatial) and 'MOUNTAIN BUILDING' (temporal).

b Describe the types of change being shown in Figure 2.2A, B, C and D. Use ideas from the geography dictionary definition of change in your answer.

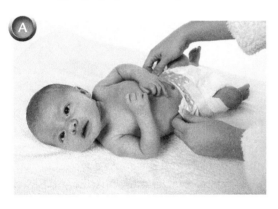
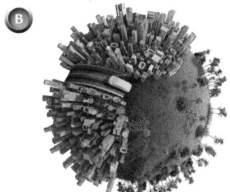

Fig. 2.2

2 a Write a short paragraph describing the scene in Figure 2.3.

b What events/things could have brought about the changes to the natural environment shown?

c Suggest the point that is being made by including a child in the picture. How does the child add to the picture?

Fig. 2.3

3 Copy Figure 2.4 and add the words 'spatial' and 'temporal' to the appropriate axis.

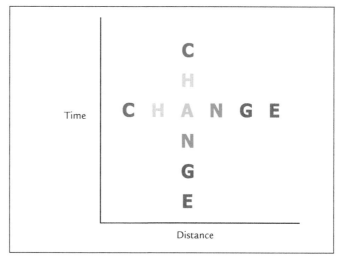

Fig. 2.4

4 Illustrate one of the three quotations about change in Table 2.1.

Education is the most powerful weapon which you can use to change the world. (Nelson Mandela)
Nothing that is can pause or stay; *The moon will wax and wain,* *The mist and cloud will turn to rain,* *The rain to mist and rain again,* *Tomorrow will be today.* (Henry Wadsworth Longfellow)
You cannot step twice into the same river, for other waters are continually flowing in. (Heraclitus, 500 BC)

Table 2.1

ISBN: 9780170215701

Weather and climate across New Zealand

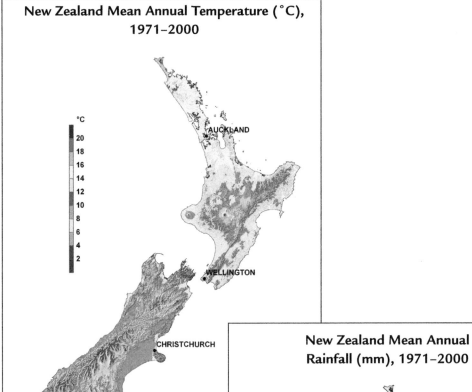

New Zealand Mean Annual Temperature (°C), 1971–2000

Fig. 2.5

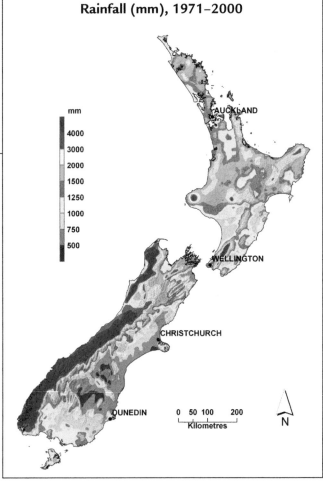

New Zealand Mean Annual Rainfall (mm), 1971–2000

Fig. 2.6

ISBN: 9780170215701

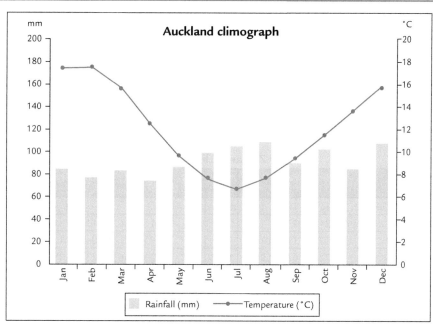

Fig. 2.7

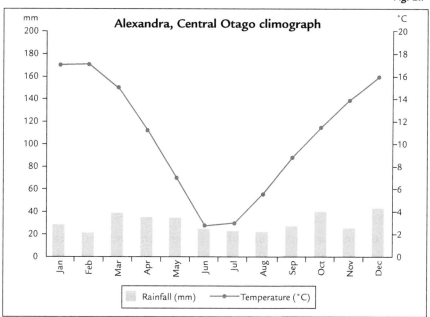

Fig. 2.8

1   a   Describe the temperature pattern shown in Figure 2.5. Use the words 'change' and 'variation' in your description.

b   Use an atlas and the information and data from Figure 2.6 to copy and complete this table.

	North Island	South Island
Wettest places		
Fairly wet		
Medium rainfall		
Fairly dry		
Driest places		

ACTIVITY

ISBN: 9780170215701

c Choose either Figure 2.7 (Auckland) or 2.8 (Alexandra) and describe the changes in temperature and rainfall that take place over the year. Refer to months and seasons and include specific statistics in your answers.

2 a Choose one of the four locations in Table 2.2. Using the statistics construct a climograph for that location (follow the style of the Auckland and Alexandra graphs).

New Zealand Climate Statistics

Location		Jan	Feb	Mar	Apr	May	Jun	Jul	Aug	Sep	Oct	Nov	Dec
Tongariro National Park	Temp (°C)	11	12	11	8	5	3	2	4	4	6	8	10
	Rain (mm)	218	226	170	241	218	264	236	211	231	264	213	267
Wellington	T	17	17	16	14	12	10	9	10	11	12	13	16
	R	72	62	92	100	117	147	136	123	100	115	99	86
Christchurch	T	17	17	15	13	9	7	6	7	10	12	14	16
	R	42	39	54	54	56	66	79	69	47	53	44	49
Milford Sound	T	15	15	13	11	8	6	5	7	8	10	12	13
	R	717	499	640	585	641	440	418	427	523	688	522	648

Table 2.2

b Read-off data from the Auckland and Alexandra climographs to produce a table of rainfall and temperature figures month by month for the two locations (follow the style of Table 2.2).

c Referring to the temperature and rainfall statistics for the four locations in Table 2.2, copy and complete the following statements.

   i _____ is the place with both the coldest summer and coldest winter temperatures.

   ii Wellington and _____ both have much more rain in winter than they do in summer.

   iii Every month _____ is the place with the highest rainfall total.

   iv One of the four places has its highest temperatures and greatest amount of rain in the same month. This place is _____ and the month is _____.

d Study Figures 2.5 and 2.6 and the statistics in Table 2.2. Which climate element (rainfall or temperature) shows the greatest change over the year?

3 a On an outline map of New Zealand locate and name the six locations referred to in Activity 2, and the place where you live.

   b Study Figures 2.5 and 2.6 and then estimate the annual average temperature and average annual rainfall for where you live.

   c Referring to Figures 2.5 to 2.8, give an example of:

      i Spatial change in weather or climate.

      ii Cyclical change in weather or climate.

ISBN: 9780170215701

Dubai and longitude 55

Dubai is one of seven emirates that together make up the United Arab Emirates (UAE). The largest of the seven emirates is Abu Dhabi, and Dubai is the second largest. It is located on the Persian Gulf coast at latitude 25°N, longitude 55°E. Dubai is located in a desert area with a hot and very dry climate. The population total is approaching 2 million and the land area is around 4000 sq km (similar in size to the Auckland city region).

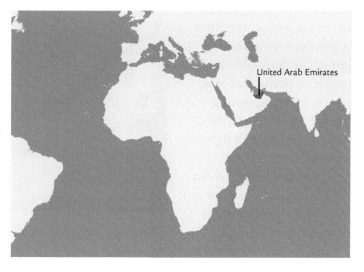

United Arab Emirates

Oil was discovered in 1966 and since then the country has changed to become a global city, a business and financial centre, international airport hub and tourist destination. The Burj Khalifa is the world's tallest building. 95% of the income of the country now comes from non-oil sources. Huge building projects, cultural festivals and 'rich' sports events (golf, tennis and horse racing) attract lots of world attention and many visitors to Dubai.

People from all over the world have moved to live and work in Dubai, some as highly paid executives for global corporations and others working as low paid service workers or labourers on construction sites. Only 15% of the population are indigenous Emirati (born in the UAE). Most of the 85% who were born overseas are of Asian origin – mainly Indian and Pakistani migrant men – who work as service workers and labourers. Males outnumber females by about three to one.

ISBN: 9780170215701

Prosperity, innovation and modernity are one side of the city image, but there is another side of stalling growth because of financial problems in the economy, coupled with increasing worker unrest due to the low pay and human rights abuses.

Questions persist about whether a city built (and reliant) on huge amounts of money and technology in a desert environment is sustainable in the long term. Getting enough fresh water and disposing of waste water are two huge problems the city faces. Dubai has the largest carbon footprint per person on the planet. In population and area Dubai may be small, but it has a reputation that reaches well beyond its size.

The Burj Khalifa.

ISBN: 9780170215701

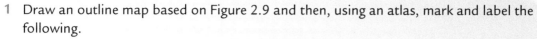

1   Draw an outline map based on Figure 2.9 and then, using an atlas, mark and label the following.

- Dubai.
- Longitude line 55° east, latitude line 25⁰ north and the tropic of Cancer.
- Desert areas (pale land colours on Figure 2.9) and wetter areas (green land colours).
- Names of the surrounding seas and oceans labelled C, G, I, M and R.
- Names of the countries along latitude 25 and longitude 55.

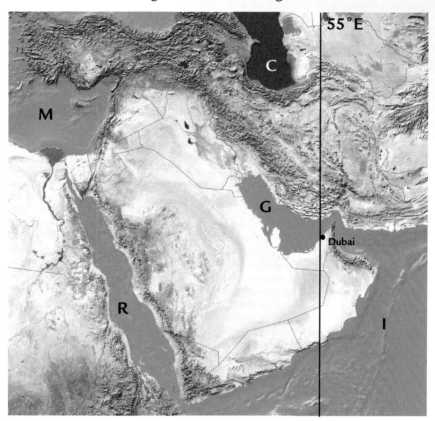

Fig. 2.9 The Middle East

2   Read the information in Table 2.3 about Dubai and study Figures 2.10 and 2.11. Write a detailed description and explanation of how Dubai has changed since 1961.

Dubai: a place transformed	
1961	2011
Looking out over the old spice market it was a place of sand tracks and narrow lanes between low sun drenched buildings. Old wind towers diverted air down into the buildings below. Donkeys, horses and camels were the main form of transport. There were rarely cars to be seen, and there was no electricity, no running water, no glass and no concrete. For most people the day started with getting water from a well. They were poor people without much hope of a better life. The tallest buildings were minarets visible across the sandy desert in the far distance. The sleepy old port was famous for pearl diving and fishing.	The city has a Manhattan-like profile. It has become the highest of high rise cities. Air-conditioned glass and concrete towers and construction cranes dominate the skyline. Malls, hotels and shops line the city streets with global names and designer brands. It is a city of huge four wheel drive vehicles, with twelve lane highways leading from the city centre out into the desert beyond. Artificial islands line the shoreline with homes and apartments waiting for millionaire buyers. The city has been transformed from the 18th to the 21st century in less than 30 years. Dubai sees itself as a model, trend-setting modern Muslim country.

Table 2.3

ISBN: 9780170215701

Fig. 2.10 Dubai in the 1960s

Fig 2.11 Dubai in 2010

ISBN: 9780170215701

3 Figures 2.12 to 2.16 show a range of images taken while travelling south along longitude 55°E. Draw a series of simple annotated sketches titled 'Landscape change in Dubai along the line of longitude 55° east'.

Fig. 2.12 At the coast: Dubai city centre by day

Fig. 2.13 At the coast: Dubai city centre at night

Fig. 2.14 At the city edge: expanding into the desert

Fig. 2.15 Inland sand desert landscape

Fig. 2.16 Inland rocky and mountain desert

ISBN: 9780170215701

Global settlement and the rise and fall of Detroit

At some point in the first decade of the 21st century a baby was born in a city hospital, or a family moved from a rural area to live in a town, which meant that for the first time in history more of the world's population was living in urban than in rural areas (Figure 2.17).

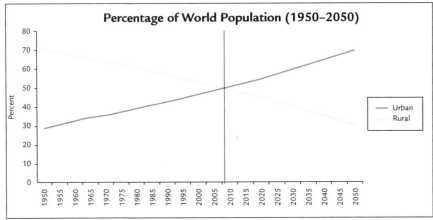

Fig. 2.17 Tipping point

The growth in the number of people living in towns and cities (urban areas) and in the size of city populations has been a feature of world geography during the last 200 years. Villages have grown into towns, and towns have grown into huge cities. Some cities have grown out of nothing, on land that was empty space only a few decades ago. (See Case study 2, page 21.)

A closer look at city populations reveals some important trends. Not all cities have been growing at the same rate. In some cities, population growth continues to accelerate, while in other cities the rate of growth has slowed (see Figure 2.18). Then there are cities that go against all these trends, and have declining populations.

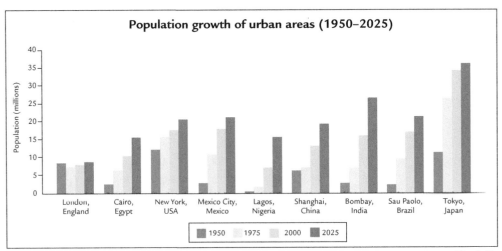

Fig. 2.18

New York and London are typical of large cities in more developed countries that grew rapidly in the 1800s and early 1900s, reached their current size mid-20th century, and have since experienced slow growth or decline. Cities in some less developed countries, such as Mexico City, grew very rapidly between 1950 and 1980, but are growing more slowly now. Many Asian and African cities, such as Bombay and Lagos, are experiencing very rapid growth now and are projected to continue at this pace.

ISBN: 9780170215701

There are another group of cities that go against this urban growth trend. These are cities that have been losing population. In some, such as New Orleans in the US and Christchurch in New Zealand, natural disasters (Hurricane Katrina and earthquakes) have been the major cause for people leaving the city. In another group of well-established cities – especially those in Europe and North America – the closure of major industries and job losses resulted in population decline over many years. Detroit, USA is one of these cities.

Detroit: Motor City in decline

Fig. 2.19 Regional location and site

Fig. 2.20 Form (design, land use and layout) of the city from CBD to suburbs

In 1913 car manufacturer Henry Ford introduced large-scale assembly line production in Detroit. The city soon became the centre of the world automobile industry. Ford, General Motors and Chrysler all set up assembly plants and headquarters in Detroit. Thousands of migrants, many of them African Americans from the South, came to find work in the car plants and linked car-part manufacturing industries. Others came from Europe. By 1950 Detroit's population was almost 2 million in total. It was the 4th largest city in the United States. It was a city of wealth, skyscrapers and smart neighbourhoods. As well as being 'Motor City' it was also famous for its range of large, grand and elaborate building styles both in downtown and suburban areas.

At the beginning of the 1950s, car factories were relocated to Detroit's periphery (outskirts of the city). The white middle-classes began to leave the inner city and settle in newly built suburban towns. Racial tensions and social inequalities within the city led to many violent urban riots. These things, plus high unemployment when car plants and associated manufacturing industries scaled back production, or closed down after the

ISBN: 9780170215701

First Model T Ford (left) and a modern Ford Mustang (right).

1970s, led to a large outflow of people – especially from old inner city neighbourhoods. Outdated downtown inner city buildings emptied of businesses and of people. Some people moved to more prosperous outer suburbs, many left the city forever.

With fewer people and less tax revenue, social services such as education, health care, public transport and even street lighting were cut back. The city had become trapped in a downward spiral of decline. Places of wealth and signs of prosperity can still be seen, but abandoned factories, apartment buildings and houses can be found all across the city. Many have been stripped bare by looters, others lie rotting or burnt out. One quarter of the urban area is classified as abandoned. In the past there were prosperous suburbs ringing the city. Even these suburbs today are suffering from recession.

Between 1950 and 2000, Detroit lost more than half of its population. Since 2000 the losses have continued: between 2000 and 2010 Detroit lost 237,000 more people, a decline of 25% from the year 2000 population total. Unemployment is close to 30% and homes can be bought for as little as $12,000 (NZ). The population total today is little over 700,000. The city now ranks as the 18th largest city in the United States.

The challenge for Detroit planners has been to prepare a blueprint for a smaller city – some run down and abandoned neighbourhoods will not survive other than as parks and green spaces, and there are plans to turn some of these areas into farmland within the urban area. The design focus is on creating a more compact urban area built around the vibrant suburban communities that have survived, and a rejuvenated downtown and waterfront where public open spaces and walkways are part of the design alongside new government and commercial buildings.

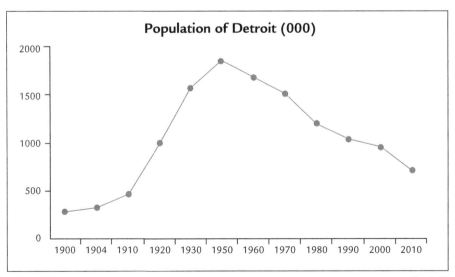

Fig. 2.21

ISBN: 9780170215701

Two faces of Detroit

Fig. 2.22 Maybe not so welcoming!

Fig. 2.23 Downtown Detroit

Fig. 2.24 An abandoned factory

Fig. 2.25 An upmarket shopping mall

Fig. 2.26 Abandoned inner city house on fire

Fig. 2.27 City edge housing

ISBN: 9780170215701

ACTIVITY

1 a   Population changes. Refer to the text and Figs 2.17 and 2.18 to copy and correctly match up the sentences in the following heads and tails table. Important terms are underlined.

Heads	Tails
1 Rural areas are country areas. Many people have …	a … the world's population live in urban areas than ever before.
2 Towns and cities are called urban areas. More of …	b … projected that 70% of the world's population will live in urban areas by 2050.
3 Villages are small towns. They are the smallest type of urban area. Many villages become part of …	c … moved away from rural areas and have migrated to live in cities.
4 Cities in Africa and South America have population totals that are increasing at a faster and …	d … decline in the number of people that live in them.
5 Many rural areas and some cities like New Orleans are having a loss of population. They have a …	e … some of the fastest growing cities on the planet.
6 London, New York and Tokyo are developed world cities. This means they are cities in wealthy …	f … cities when cities grow in area and population.
7 Less developed countries are the poorer countries. Although they are poor they contain …	g … countries. These cities have large population totals but their rates of growth have slowed.
8 A population projection is like an estimate. It is a prediction based on past and present trends. It is …	h … faster rate. This is called accelerating growth.

   b   Refer to Figure 2.17 to find at what date the 'tipping point' took place.

   c   What is the change that happened at this tipping point?

   d   Construct three pie graphs to show the world's urban and rural population make-up for 1950, 2000 and 2050.

   e   Refer to Figure 2.18 and complete the following table to show the largest and smallest of nine cities for each of the years 1950, 1975, 2000 and 2025. Add your own prediction for the largest and smallest in 2050 and justify (give reasons).

Year	Largest	Smallest
1950	New York	
1975		
2000		Lagos
2025 (projected)		
2050 (prediction)		

2   Find Detroit on an atlas map of North America. Draw a simple sketch map to show its location within the US. Include on your map the Great Lakes, the US/Canada border, the state in which Detroit is located, two big cities that are near to it and a line of latitude and a line of longitude for the city.

ISBN: 9780170215701

3 a Refer to Figures 2.19 and 2.20 to describe the site and form of Detroit.

  b Read 'Detroit: Motor City in decline', then make a copy of Figure 2.21 (page 28) and annotate with information that describes and explains the population trends on the graph.

  c Construct three bar graphs to show the population total of Detroit in 1900, 1950 and 2000.

  d What could influence the population growth or decline of Detroit between now and 2050?

4 Study the two faces of present day Detroit in Figs. 2.22–2.27 and make a half page copy of the circular diagram below. In each of the six outer segments of the diagram write four words that describe features shown in each of the photos. Colour-code each segment as either 'decay' or 'prosperous'. A start has been made for you.

KEY

 decay

prosperous

ISBN: 9780170215701

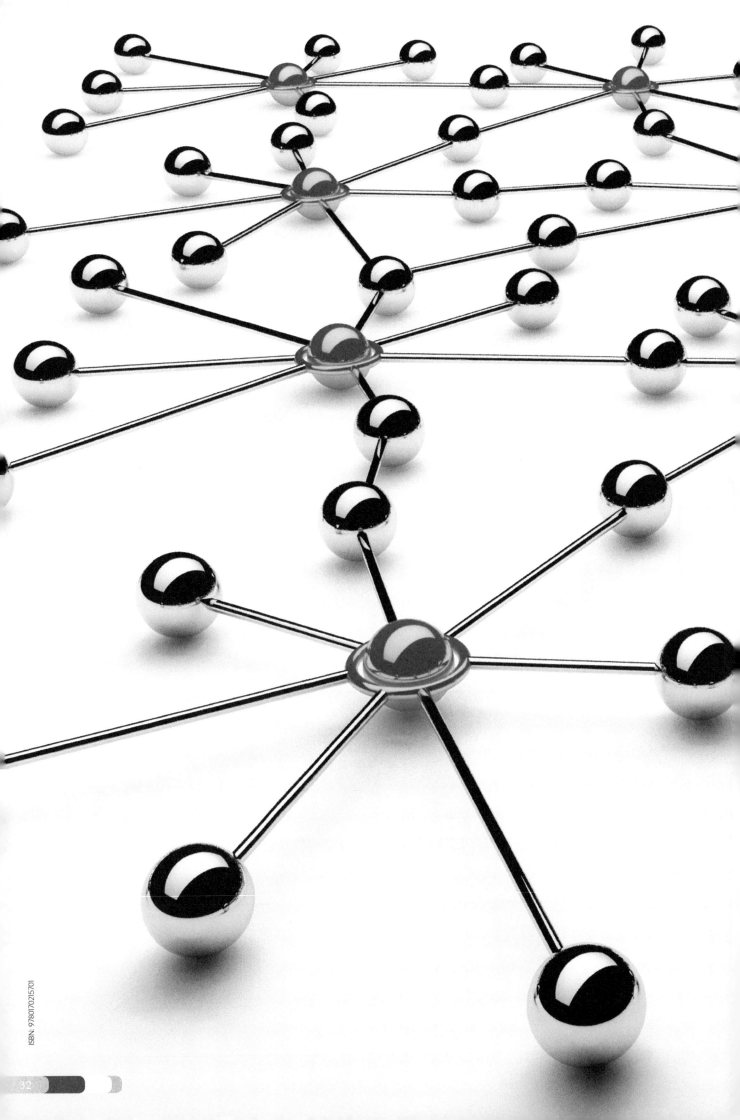

3

Connections

Globalisation

A connected world: local connections and global connections.

ISBN: 9780170215701

ISBN: 9780170215701

GEOGRAPHY DICTIONARY

Connections involve linkages between places, and between people and places. Connections have effects on places and people, and can bring about change. Connections are an important part of the geography of places.

We live in a world more connected today than it has ever been in the past. Across the globe, at any time, there are movements taking place of people and products. Originally these movements were physical, over land and sea before air travel added another way that people and places were connected. Now we have a world connected electronically, and connections between people and places can take place across the globe instantaneously. These connections are unseen until the TV, Internet or phone are switched on. People and places are affected by these connections.

Connections and movements are a part of global geography: they are taking place now. They also help explain global geography because the world as it is today has been influenced by these connections and movements.

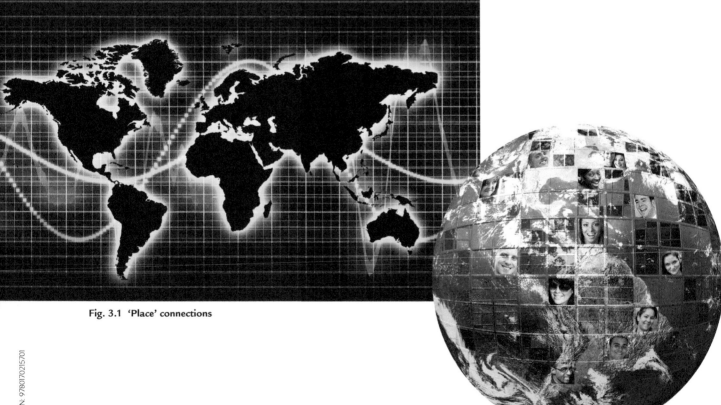

Fig. 3.1 'Place' connections

Fig. 3.2 'People' connections

Globalisation

Globalisation is one of many sub-concepts that fall under the Connections heading. In geography today, globalisation is so important that it could be classed as a major concept in its right – similar to the way 'connections' and 'interaction' are viewed as a major concepts.

Globalisation involves connections between places across the world and the way places have become more and more interdependent and interconnected. The connections are faster, cheaper and more efficient than ever before. Globalisation affects the clothes we wear, the music we listen to, the food we eat, and the environment we live in. Our world is a global world: events, activities and decisions in one part of the world can have consequences for people and communities in other parts.

Five types of globalisation

1 Movement of people across the globe, in business, as tourists or refugees, for example.
2 Instant distribution of media via TV, mobile phones, the Internet, and on radio.

Moving to escape floods in Pakistan.

Tourists in Hong Kong.

3 Flow of money internationally as companies and people invest in other countries. Ford cars, Nike sports gear and McDonald's food outlets, for instance, are found the world over; overseas companies buy New Zealand farms.
4 Movement of technologies around the globe, such as new medical drugs or glasshouse food production.
5 The global spread of ideas and thinking. Mobile and electronic devices allow organisations, such as Greenpeace and World Vision, to communicate their campaigns and activities to protect the environment and for social justice, to a global audience.

ISBN: 9780170215701

ACTIVITY

1 a List in alphabetical order the 12 'connection' words from Figure 3.3.

				I	N	T	E	R	A	C	T	I	O	N						
								F	L	O	W	S								
				M	O	V	E	M	E	N	T									
					I	N	T	E	R	N	E	T								
						T	R	A	V	E	L									
C	A	U	S	E	&	E	F	F	E	C	T									
										T	R	A	D	E						
									M	I	G	R	A	T	I	O	N			
								G	L	O	B	A	L	I	S	A	T	I	O	N
	I	N	T	E	R	D	E	P	E	N	D	E	N	C	E					
						L	I	N	K	S										

Fig. 3.3

b For each connection word choose one of the icons from Figure 3.4 to represent it, and give an example from geography (connected with places, the environment or people) of this word in action. There are several geography statements to select from on page 37. The first has been started for you.

Word	Icon	Example of the word in action
1 Cause & effect	⤓	A big earthquake off the coast of Japan caused huge tsunami waves. These waves destroyed buildings, flooded the land and killed many people.
2 Connections		

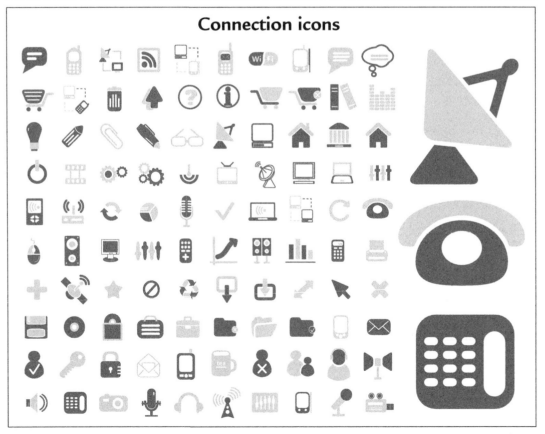

Connection icons

Fig. 3.4

ISBN: 9780170215701

Example ideas to choose from

- Exports of New Zealand logs to China have increased, while imported Chinese-made flat-screen TVs dominate the New Zealand market.

- New and improved rail tracks and modern carriages have led to increased use of rail by commuters in Wellington and Auckland.

- A big earthquake off the coast of Japan caused huge tsunami waves. These waves destroyed buildings, flooded the land and killed many people.

- There are strong connections between Australia and New Zealand in tourism. Many Australians come for skiing holidays in Queenstown, while New Zealanders frequently fly to the Queensland Gold Coast for beach vacations.

- Within Europe, people can travel freely across the borders of European Union countries to live and find work outside their home country.

- In New Zealand the movement of people away from Auckland to smaller cities like Tauranga has become a feature of internal migration.

- Heavy bursts of rain saturated the land, caused floods and massive landslips on farmland near the Hawke's Bay coast.

- The Internet brings the outside world into people's homes. Many people now work from home, place orders for goods online, make overseas investments or arrange overseas travel without ever leaving the house to visit an office or shop.

- Sports events like the Olympics and soccer and rugby World Cups transform cities, and people criss-cross the world to get to these events. Satellites beam the events into homes and communities in every continent.

- People escaping from famine and fighting in Somalia, Ethiopia and Kenya move into refugee camps in Kenya. Medical and food aid is delivered to the camps from other parts of the world.

- People and the atmosphere depend on one another. Rain, sunshine and warmth allow for human survival, but global warming has shown that the atmosphere is being greatly changed by human actions.

- Cities worldwide rely on an underground network of pipes and cables to provide them with the power, communications, water and sewage disposal they need to function.

2   Make two quick drawings of Figures 3.1 and 3.2, then choose the word that you think each drawing best illustrates from the list of terms in Figure 3.3. Write this word across each sketch you have made as if the term was a 'passport approved' stamp on the drawing. Beneath each sketch give a reason for your choice of term.

ISBN: 9780170215701

3　Figure 3.5 shows 'Home' connections: the blue spheres represent where people live (their homes), and the black spheres represent other places. The lines show connections between people's homes and these other places. Imagine your home is one of the blue spheres, suggest what the black spheres and lines could be. Answer either in paragraph form or label a copy of the diagram. Include real place names in your answer. These should be the names of places your home is connected with, for example where your relations live, where you go to school, where you went on vacation or where your Facebook friends live. The lines could represent how you travel to or connect with these places.

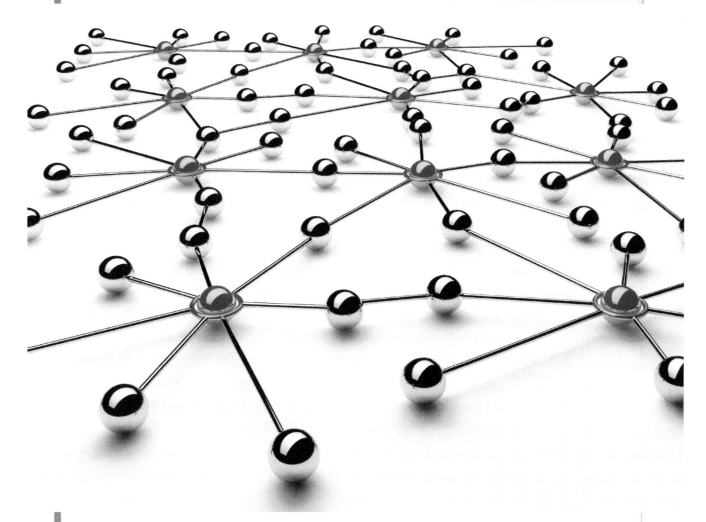

Fig. 3.5　'Home' connections

ISBN: 9780170215701

Facebook: connections across the world

'Facebook before tap water'

In a two-room shanty with no running water in the Indian city of Mumbai, Darshana Verma makes tea on a small stove. On a bench nearby, her 18-year-old daughter Shanti messages Facebook friends on the keypad of her Nokia smartphone.

'This is the Internet age,' said the 36-year-old domestic helper, who spent more than half her NZ$357 monthly income on Samsung electronics and Nokia mobile phones for her children. 'Facebook is there, all these things happen there now. They make friends, maybe they can even find jobs there.'

For Verma, who never learned to use a computer and saved for ten months to buy her eldest daughter's phone, that gives her children an opportunity she didn't have. 'What I don't know about Facebook and

the Internet, they need to know about,' she said. 'It is worth the expense.'

In China, the world's most populous nation, Facebook is banned and blocked. China – the world's largest Internet market with more than 450 million users – has also banned pornography, gambling and content critical of the ruling Communist Party from the Internet. 'Facebook has chosen to focus on open markets, rather than markets like China where there's censorship and control,' said Foong King Yew, vice-president of research at Gartner in Singapore. 'India's the biggest of those. It's rapidly growing. It's an untapped market.'

Fig. 3.6

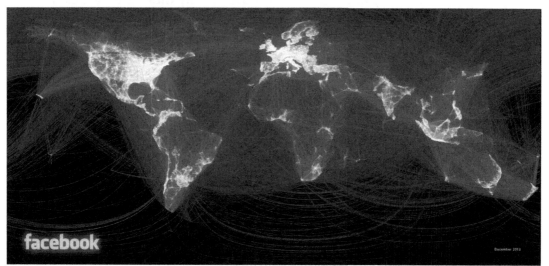

Fig. 3.7 Visualising friendships

ISBN: 9780170215701

The global spread of Facebook is represented in Figure 3.7. At first sight this social network appears to be used all across the globe, but a closer look at the map reveals that this is not the case. There are gaps. In large parts of Brazil, Russia, central Asia, China, Africa and Latin America there seems to be little or no activity. The map was constructed by Paul Butler, who was interested in the locations of his online friends so decided to create a visualisation of Facebook connections. How local are our friends? Where are the highest concentration of friendships? How do political and other geographical boundaries affect them?

Butler started by using a sample of ten million friend pairs, matched them with where they lived, and then mapped that data using the longitude and latitude of each location. Each place was represented on the map with a dot, then lines were drawn to show the connections between two dots (two friends). Areas with large concentrations of Facebook users show as white areas on the map. Where traffic is heavy between two areas the connection lines merge, overlap and blur together creating purple shading on the map.

Population distribution and wealth influence the pattern. Since Facebook is people-driven, it is no surprise that Facebook use is higher where lots of wealthy people live. Europe and eastern USA fall into this category. Many regions of low Facebook use are areas of low population density, for example the interior of Australia. Across the Sahara Desert in Africa the lack of network coverage, as well as low population density, also result in low use of Facebook.

There are other areas, however, where the lack of Facebook use is surprising. Brazilians seem to prefer Bebo.com, while Vkontakte.ru is hugely popular in Russia. China has three major social networking websites, with Renren.com being the most popular. Chinese users seem to be more focused on the gaming aspect of social networks.

ACTIVITY

1 a Who is Darshana Verma?

   b What evidence is there that she regards the Internet, Facebook and the mobile phone as being important?

   c What reasons does she give for valuing the electronic communications devices so highly?

2 a Make a copy of the table that follows and then, using Figure 3.7 and an atlas to guide you, write correct place names in the spaces provided.

Area of the world	Name of country, region or city with *high* Facebook use	Name of country, region or city with *low* Facebook use
North America		
South America		
Africa		
Europe		
Middle East		
Asia		
Oceania		

   b Identify and describe patterns shown on the map in Figure 3.7. Include examples of place names to support your answer.

   c Give an explanation for the patterns.

   d Which of the five types of globalisation referred to on page 35 does the Facebook case study most relate to? Justify your answer.

ISBN: 9780170215701

CASE STUDY 5

Wrapped up in globalisation: Fair Trade chocolate

Trade Aid advertising:

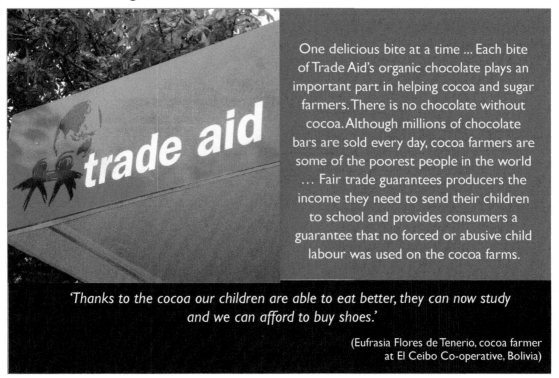

One delicious bite at a time ... Each bite of Trade Aid's organic chocolate plays an important part in helping cocoa and sugar farmers. There is no chocolate without cocoa. Although millions of chocolate bars are sold every day, cocoa farmers are some of the poorest people in the world ... Fair trade guarantees producers the income they need to send their children to school and provides consumers a guarantee that no forced or abusive child labour was used on the cocoa farms.

'Thanks to the cocoa our children are able to eat better, they can now study and we can afford to buy shoes.'

(Eufrasia Flores de Tenerio, cocoa farmer at El Ceibo Co-operative, Bolivia)

Trade Aid shops are found in main cities and towns all over New Zealand. They sell a range of food and handicraft items that are sourced from developing nations. The organisation, however, is about more than just shops. It is a not for profit organisation made up of several parts involving work as a retailer, importer, wholesaler and development agency.

Trade Aid is a member of the World Fair Trade Organisation (WFTO), an international body of organisations committed to Fair Trade. This means paying people from outside of New Zealand a fair price for what they produce, making sure that working conditions for these people are satisfactory and that their rights are protected. Trade Aid works with suppliers mainly in developing nations in Africa, Asia, Latin America and the Pacific, and aims to improve the lives of disadvantaged producers around the world. Trade Aid often works with community groups and co-operatives in these countries. It buys their produce and makes sure the 'growers and makers' receive a fair price for their products.

Fair Trade chocolate is sold widely in New Zealand. It is high-quality chocolate that is sold at a 'fair price'. This product shows globalisation at work. Similar brands of Fair Trade chocolate are sold in Europe, Australia and North America including some made by global multinationals like Cadbury and Nestle. The New Zealand chocolate is produced from raw materials grown in different tropical countries, is manufactured in Europe, then packaged and imported into New Zealand to a distribution centre before being sent across the country for sale (Figure 3.8).

Fair Trade chocolate has two other features that make it good for the environment: the raw ingredients are organically produced, and the packaging is made from recycled materials.

ISBN: 9780170215701

Fair Trade chocolate from raw materials to the retailer

Belgium and Switzerland:
Chocolate manufacture
and packaging

Christchurch Distribution
Centre

Paraguay and
Philippines: Sugar
production

Dominican
Republic, Bolivia,
Peru and Ghana:
Cocoa production

Trade Aid shops around
New Zealand

Fig. 3.8

1   Design an advert for Fair Trade products aimed at increasing
    sales to high school students.

2   a   What reasons can you think of why
        major chocolate manufacturers like
        Whittakers, Nestle and Cadbury
        would want to produce and
        market some Fair Trade
        products in their range?

    b   How does Fair Trade
        chocolate illustrate the
        concept of globalisation? Refer to
        the five types of globalisation listed
        previously on page 35.

3   Show the information from Figure 3.8 on a map
    of the world.

ISBN: 9780170215701

Global corporations and global products: dine on a Subway

Many companies operate across the globe. These are called multinational companies. They have their headquarters in one country (usually the USA) but manufacture and sell in many other countries. Many of these companies are well known names. Fonterra (dairy products) is New Zealand's best known global company. Fisher and Paykel (whiteware) and Rakon (electronics: design and manufacture of frequency control products used in GPS units and mobile phones) are other New Zealand examples. These companies are tiny in comparison with global giants such as McDonalds, Microsoft, Google and Toyota.

Fig. 3.9 Subway restaurant in Belo Horizonte, Brazil

Subway has become the largest fast food chain in the world, surpassing McDonalds and dwarfing Burger King. With its 'Eat fresh' advertising slogan, Subway now operates in 98 countries. It is a global giant. Subway has earned a reputation for opening in some unlikely places such as a United States air base in Bagram, Afghanistan, in a church in Buffalo (New York state), and even on a riverboat in Germany. Subway has captured a lot of publicity by attaching a Subway store to the scaffolding of One World Trade in Manhattan, New York, the new skyscraper being built where the Twin Towers used to stand. Built inside a pod (like a shipping container), which rises as the building gets ever taller, it supplies Subway sandwiches to the steelworkers on the construction project.

Back in 1965, Fred DeLuca was searching for a way to help pay for his training as a doctor. A family friend (Dr Peter Buck) suggested he open a 'submarine sandwich shop'. With a loan of $1000 Dr Buck offered to become Fred's partner, and a business relationship was forged that changed the landscape of the fast food industry. The first store was opened that year in Bridgeport, Connecticut, USA. By 1974, the pair owned and operated 16 submarine sandwich shops.

To speed up the growth of the company they began franchising. This launched the Subway brand into a period of rapid growth. The first overseas Subway store was opened

ISBN: 9780170215701

in Bahrain in 1984. The first New Zealand store opened in Parnell, Auckland in 1995. Now Subway can be found in more than 35,000 locations around the world (Table 3.1). 'We've become the leading choice for people seeking quick, nutritious meals that the whole family can enjoy' said Fred. He sees further growth as the company 'continues his passion for delighting customers by serving fresh, delicious, made-to-order sandwiches'.

Subway's main office is in Milford, Connecticut. Regional centres support Subway's international operations: the regional offices for European franchises are located in Amsterdam, Netherlands; the Australia and New Zealand locations are supported from Brisbane, Australia; the Middle Eastern locations from offices located in Beirut, Lebanon; the Asian locations from Singapore and India; and the Latin America support centre is in Miami, Florida, USA.

Glocalisation

This is a term used when a standardised product is adapted to suit local tastes and culture. Subway illustrates glocalisation. It has opened kosher restaurants in areas of the US with large Jewish populations, using menus containing no pork-based products and using soy-based cheese. Subway restaurants in Muslim countries offer a halal menu to its local consumers. Non local customers can purchase from an alternative menu. There are also two halal Subway restaurants in the US that do the same, three in Canada, and over 100 in the United Kingdom. Subway is planning to open more of these restaurants.

ACTIVITY

1   Explain, using examples, the meaning of the term 'multinational company'.

2   Research: find and create a full page collage of the logos of multinational companies.

3   Draw a star diagram that gives reasons for the success and growth of Subway.

4   How does the Subway story provide examples of the concepts of 'connection', 'globalisation' and 'glocalisation'?

5   a   Show the data in Table 3.1 on a world map using either proportional symbols or choropleth shading.

     b   Describe and suggest an explanation for the pattern that the map shows.

ISBN: 9780170215701

Subway restaurants around the world

Country	Number of Subway stores	Country	Number of Subway stores
Afghanistan	2	Liechtenstein	1
Antigua and Barbuda	2	Luxembourg	11
Argentina	22	Macau	2
Aruba	7	Malaysia	80
Australia	1283	Marshall Islands	1
Austria	7	Martinique	2
Bahamas	4	Mexico	533
Bahrain	11	Netherlands	90
Barbados	1	Netherlands Bes Islands	1
Belgium	17	New Zealand	233
Bolivia	10	Nicaragua	11
Brazil	683	Northern Mariana Islands	3
Bulgaria	17	Norway	14
Canada	2676	Oman	10
Cayman Islands	5	Pakistan	29
Chile	21	Panama	39
China	228	Peru	6
Colombia	33	Philippines	13
Costa Rica	40	Poland	44
Curacao	8	Portugal	7
Czech Republic	8	Puerto Rico	206
Denmark	4	Qatar	15
Dominica	2	Russian Federation	269
Ecuador	6	Saint Kitts and Nevis	1
Egypt	4	Saint Lucia	2
El Salvador	26	Saint Vincent and Grenadines	1
Finland	89	Saudi Arabia	44
France	288	Singapore	92
Germany	614	Sint Maarten	6
Gibraltar	1	Slovakia	9
Greece	4	South Africa	15
Grenada	3	South Korea	42
Guam	13	Spain	41
Guatemala	35	St Martin	1
Honduras	19	Sweden	90
Hong Kong	21	Switzerland	8
Hungary	10	Taiwan	126
Iceland	19	Tanzania	6
India	230	Thailand	37
Iraq	5	Trinidad and Tobago	38
Ireland	103	Turkey	15
Isle of Man	2	United Arab Emirates	109
Israel	8	United Kingdom	1415
Italy	10	United States	24544
Jamaica	3	Uruguay	2
Japan	275	Venezuela	160
Jordan	7	Vietnam	1
Kuwait	48	Virgin Islands (US)	8
Lebanon	4	Zambia	5

Table 3.1

Environments

4

Places
Regions
Tūrangawaewae

ISBN: 9780170215701

Port Underwood and Opihi Bay in
Marlborough Sounds: an environment
shaped by nature and people.

ISBN: 9780170215701

GEOGRAPHY DICTIONARY

Environments are surroundings. They are the conditions found in a place or the conditions that surround us. Environments may be natural and/or cultural. They have particular <u>characteristics</u> and features which can be the result of natural and/or cultural <u>processes</u>. The particular characteristics of an environment can be similar or different to those of another environment.

The environment is made up of the things (features or elements) that <u>surround</u> us. These elements may be natural and/or cultural. If you were going on a trip to a mountainous area, the trip could be described as a trip into a mountainous environment. A school located in a farming area could be described as a school in a <u>rural</u> environment.

An environment is made up of many parts (many features or elements). If the features are all <u>natural</u> then the environment is described a 'natural environment'. If the features are all cultural then the environment is a '<u>cultural</u> environment'. Often environments are made up of a combination of natural and cultural features (Figure 4.1).

Knowing the origin of a word can help us to understand what it means. Environment comes from the word '<u>environs</u>', meaning 'surroundings'. Environs has another meaning, to do with a 'circle' or 'ring' around things. Think of environment as being about the things that surround us. It means the conditions that exist inside a circle or ring around us or around a place.

Environment has connections with other geography concepts like places, areas and regions. Often these words can be <u>substituted</u> one for another without altering the meaning. For example, desert environment, desert region or desert area all carry the same or very similar meaning. Environment is important to geography because geography is about places, and is concerned with describing the environments of places, <u>regions</u> or areas. Geography will frequently ask two questions about a place:

1   What is the place like? The answer would describe the parts that make up the environment. Putting all these parts together provides a picture of the place.

2   Why is the place like that? The answer would try to explain how parts of the environment got into their <u>current</u> state.

Fig. 4.1 The global environment has a mixture of natural and cultural features

Figure 4.2 Mackinnon Pass on the Milford Track, Fiordland, New Zealand

Places and Regions: these two concepts are closely related.

A 'place' is an area of space for example a building, a beach, a town, a forest, a farm, or a district.

A 'region' is an area of the earth's surface that has common features within it. These features make it different from areas that surround it. The features can be natural or cultural. A mountainous region like the Mackinnon Pass shown in Figure 4.2 has high relief, steep slopes and a cold climate. This is a region made up of distinctive natural features. These are features that make it different from surrounding lower regions.

Another type of region are those like Waikato, Hawke's Bay, Canterbury and Otago. These are regions with their own local councils. These regions have been created by people and they are different one from another because they are controlled and governed by different local bodies.

ACTIVITY

1 a   The word 'environment' is commonly used. It is used in each sentence in Table 4.1. Describe or list of the type of features you would expect to find in each environment that is mentioned. For example in sentence (i) the difficult environment to grow up in could be a crowded house. What other things could make the environment difficult to grow up in?

In sentence (ii) having lots of computer and Internet access could be one example. Add more features that would make the environment co-operative as well as high tech.

In sentence (iii) give examples of how the environment could be damaged.

i	'The children grew up in a difficult environment.'
ii	'The school focuses on developing a co-operative and high tech learning environment.'
iii	'Industrial and farm development is causing widespread damage to the environment.'

Table 4.1 Examples of the word 'environment'

b   There are ten words underlined in the text on page 48. Make a list of these and give the meaning of each. Use a dictionary for words you are unsure of.

2   Make a large copy of the circle on page 50 to represent the world as shown in Figure 4.1. Write examples into the circle of features of both the natural and cultural environment of the world that you can see in Figure 4.1. Add other examples of environmental features of the world that you can think of.

ISBN: 9780170215701

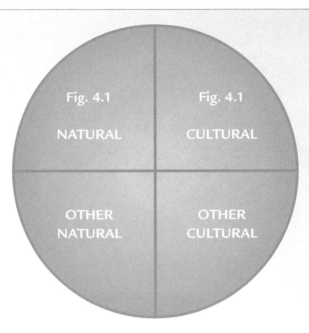

3 a Describe the environment shown in Figure 4.2 in a 30–40 word paragraph.

b Is this environment natural or cultural? Give a reason for your answer.

c If you could ask an expert three questions to confirm your answer to b, what would you ask?

ISBN: 9780170215701

CASE STUDY 7

My place

'Looking out of the kitchen window I can see down to the beach and ocean. About a kilometre away. In summer it is packed with people. Today in winter there are only seagulls. Close to shore is a small island. In summer I kayak there with my friends. Dad says that go beyond the island and the next piece of land you will see will be South America. Between my house and the beach is the road where I catch the school bus each day. It's about 40 minutes drive to school and town. Much quicker now since the council sealed the road last year. The road ends down by the beach where there is the store, a motorcamp and a few baches. A few of these are now rented out all year round. This winter after heavy rain the small river that runs down from the hill to the shore burst its banks and flooded the motorcamp. Our house has a small garden with a few sheds nearby for the tractor and other farm equipment. There are also a few small fenced paddocks near the house and sheds. There are some sheep in them waiting to go to the saleyards tomorrow when the stock truck comes. Other than that all the land around is farmland covered with grass – green in winter but yellow and dried off for much of the summer. We run sheep and a few beef cattle on the land. It is enough to give us a good income. Behind the house the road snakes up the hill where the power line comes down the ridge to our house and away down to the beach. In summer the only bit of green you can see is the pine plantation across the top of the ridge and the willow trees alongside the river.'

Fig. 4.3 From a student's written description

Tūrangawaewae

Tūrangawaewae is one of the most well-known and powerful Māori concepts and normally translates as 'a place to stand'. Tūranga means 'standing place' and waewae means 'feet'.

Tūrangawaewae are places where Māori feel especially empowered and connected. It is home. A place where you belong and are understood. Māori can describe their tūrangawaewae as 'our foundation, our place in the world, our home'.

Tāwhiao, the second Māori king, referred to tūrangawaewae in a saying that referred to three places:

> Ko Arekahānara tōku haona kaha
> Ko Kemureti tōku oko horoi
> Ko Ngāruawāhia tōku tūrangawaewae

> Alexandra [*Pirongia township*] will ever be a symbol of my strength of character,
> Cambridge a symbol of my wash bowl of sorrow,
> And Ngāruawāhia my footstool.

Pepeha (tribal sayings)

Tūrangawaewae can include other places as well. Many tribes identify themselves in terms of their mountains, waterways and important ancestors. In this way the idea of tūrangawaewae is broadened into a region located within a wider world.

When Tūwharetoa identify themselves, for example, they might say:

> Ko Tongariro te maunga
> Ko Taupō te moana
> Ko Tūwharetoa te iwi

> Tongariro is the mountain,
> Taupō is the sea,
> Tūwharetoa is the tribe.

Many people both in New Zealand and in other countries who are not Māori feel a strong connection with particular places – often this is the place they were born or where their ancestors came from. These are their 'Special Places'. Although they may not live in these special places their feeling of connection with them is strong. Some people describe their heart being in these places even though they don't live there. The concept of 'tūrangawaewae' could be described as being universal.

ACTIVITY

1 a Read the student's description in Figure 4.3 and draw a map of the area based on the text.

   b What type of environment is this? Write a short paragraph and include 'natural', 'cultural', 'rural' and 'coastal' in your answer.

2 In what ways are 'tūrangawaewae' and 'pepeha' connected with geography? What makes them Māori concepts that are linked to geography?

3 Choose a place that is special for you: 'Your place'. This could be the place you were born and brought up, a place where your ancestors came from or a place you like to visit. Complete these two tasks for 'Your place'.

   i Writing: either write a description of your place or write a pepeha.

   ii Drawing: either draw a map of your place or sketch it.

ISBN: 9780170215701

ISBN: 9780170215701

CASE STUDY 8

Yosemite National Park

Yosemite National Park is in the United States and located in the far east of the state of California. It takes about six hours to drive there from Los Angeles. It is a World Heritage park in the Sierra Nevada mountains. Yosemite is famous for its spectacular, vertical granite cliffs, waterfalls, clear streams, Giant Sequoia groves and biological diversity. Almost 95% of the park is designated wilderness.

The landforms of the park have been shaped by river and glacial erosive processes wearing away the hard granite rock. Deep narrow canyons are the most famous of the landform features. Five vegetation zones can be found in the park, which change according to height above sea level. On the lower slopes is chaparral and oak woodland, while small alpine plants dominate at high altitude. Above these there is bare rock swept regularly by icy winds and snowfall.

Marmot sitting on granite rock.

Yosemite was home for the Ahwahneechee and other indigenous people for generations before Europeans arrived in the mid-1800s. Completion of the Yosemite Valley railroad from Merced to El Portal in 1907 made access easier, and now 3.5 million people enter the gates to explore the park each year. The park has a network of walking trails, many of which can be accessed by car or shuttle bus.

Yosemite Waterfalls and chapel.

ACTIVITY

1 Make a sketch of Figure 4.4 and annotate it to emphasise features of the natural environment.

Fig. 4.4 Yosemite Valley

2 What are some of the cultural influences on the environment evident in the case study?

3 The Yosemite National Park sign (Figure 4.5) shows some features of the park and is set on a base made of granite and includes the US National Park Service logo (see inset). This logo is shaded to look as if it has been carved out of wood or rock. The elements on the logo represent the major features of the national park system: the Sequoia tree and bison represent vegetation and wildlife; the mountains and water represent scenic and recreational values; and the arrowhead shape represents historical and archaeological values. Design an entrance sign for your town, community or favourite place ('Your place').

US National Park Service logo.

Fig. 4.5 Yosemite National Park entrance sign

ISBN: 9780170215701

Different places, different environments: Antarctica and Sahara Desert

Natural environments

Antarctica and the Sahara Desert are two examples of parts of the world where nature dominates. The environments of both places are largely natural. Although these places show how natural environments can be very different from one another, there are surprising similarities.

Antarctica

Antarctica: snow, glaciers, ice flows and penguins.

Antarctica is the driest, coldest and windiest continent on Earth

- Inland the terrain is high and mountainous. Here temperatures rise to no more than -30°C in summer and fall to below -80°C in winter. At the coast temperatures are higher, but rarely above freezing.
- All year round there are strong and persistent winds. Wind speeds of over 100 km/hour are common, and gusts of over 200 km/hour have been recorded.
- Most precipitation falls as snow or ice crystals, although during summer rain can occur near the coast. The windy conditions make it difficult to accurately measure snowfall. It is estimated that the annual snowfall over the continent is equal to about 150 mm of water. Inland areas are much drier. (By comparison, Auckland and Wellington get around 1200 mm each year, and Christchurch gets 660 mm.)
- Antarctica can be described as a 'freezing desert'.

New Zealand has a particular interest in Antarctica because of Scott Base, a scientific research centre.

ISBN: 9780170215701

*Scott Base (NZ),
Antarctica.*

	Jan	Feb	Mar	Apr	May	Jun	Jul	Aug	Sep	Oct	Nov	Dec	Annual mean
Average daily temp (°C)	- 2.9	- 9.5	- 18.2	- 20.7	- 21.7	- 23	- 25.7	- 26.1	- 24.6	- 18.9	- 9.7	- 3.4	**- 16.9**
Mean daily max (°C)	- 0.2	- 6.3	- 14	- 17.4	- 19	- 19.1	- 21.7	- 22.8	- 20.8	- 15.5	- 6.7	- 0.8	**- 13.5**
Mean daily min (°C)	- 5.5	- 11.6	- 21.1	- 24.9	- 27.1	- 27.3	- 30.1	- 31.8	- 29.4	- 23.4	- 12.7	- 6	**- 20.6**
Mean monthly rainfall (mm)	15	21.2	24.1	18.4	23.7	24.9	15.6	11.3	11.8	9.7	9.5	15.7	**Annual total 202.5**

Table 4.2 Climate data for McMurdo Station
(latitude 77.88°S; longitude 166.73°E; about 24 m above sea level; 3 km from Scott Base)

ISBN: 9780170215701

Sahara Desert

Tadrart, southern Algeria: the Sahara Desert is a landscape of dunes, rocky outcrops and scattered trees.

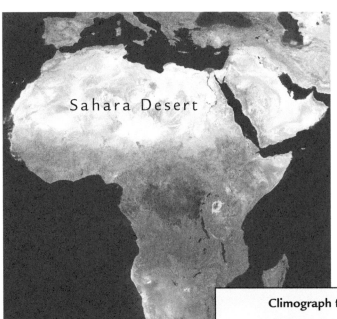

The Sahara Desert covers most of northern Africa.

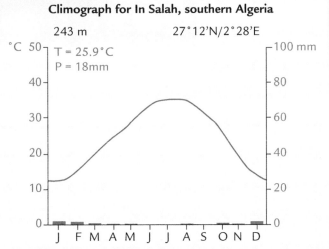

Climograph for In Salah, southern Algeria

243 m   27°12'N/2°28'E

°C 50   T = 25.9°C   100 mm
         P = 18mm

Fig. 4.6

ISBN: 9780170215701

ACTIVITY

1 a Refer to the information about Antarctica and the Sahara Desert. Make a list of ten adjectives that you could use to describe each place. Some words could be repeated in both lists – your choice. Try to include at least three words of your own in each list.

Some words could include:

ATTRACTIVE	BORING	UGLY	BEAUTIFUL	PEACEFUL
COLOURFUL	THREATENING	REMOTE	QUIET	GHOSTLY
ISOLATED	BLEAK	HOT	SHIVERING	NATURAL
VARIED	WELCOMING	VAST	EMPTY	SHIMMERING

b Construct a climograph for McMurdo Station (US) using the statistics for average daily temperature and mean monthly precipitation in Table 4.2. Construct the graph so that it can be compared with the graph for In Salah in Figure 4.6. (Remember to plan for the temperature being below freezing on your graph.)

c In what ways are the environment features of Antarctica and the Sahara very different? And what is the surprising similarity? (Hint: desert)

Other environments

2 a How have people shaped or 'made' each of the environments in Figure 4.7 A, B and C?

ISBN: 9780170215701

Fig. 4.7 Cultural environments: contrasting landscapes: farming and mining

b Imagine you lived in the type of future urban environment shown in Figure 4.8. Write a tweet or blog covering your experiences of living in this city for a week.

Fig. 4.8 Living in an urban environment: central city and outer suburbs

3 a Draw and annotate a sketch of the aerial photo in Figure 4.9. Use one colour to
identify natural features and a different colour to identify other features.
Note: the dark bands in the sea near the shore are mussel farms. Title the sketch
'Marlborough Sounds – an environment shaped by nature and by people'.

Fig. 4.9 Oblique aerial photo of the Marlborough Sounds

b Create an art-like image of a natural environment (for example Antarctica,
the Sahara Desert or the Southern Alps) following the style used for an
urban environment in Fig. 4.10.

Fig. 4.10 Art-like image of London:
urban cultural environment

ISBN: 9780170215701

Interaction

5

Sea-land interaction at the coast.

ISBN: 9780170215701

GEOGRAPHY DICTIONARY

Interaction involves elements (parts) of an environment affecting each other and being linked together. Interaction incorporates movement, flows, connections, links and interrelationships. Landscapes are the visible outcome of interactions. Interaction can bring about environmental change.

Interaction involves connections between things. A two-way effect is the essential part of the concept of interaction. Things need to have an effect upon one another for it to be interaction. 'A' having an effect on 'B' is a connection, but only if 'B' also has an effect on 'A' does it become interaction. For interaction to be a part of a geographic study the result of the interaction must impact on places, or places themselves must be involved in the interaction – there needs to be a spatial angle to the interaction.

'Things' can be called 'elements'. Geography elements are place-related things (they can also be called 'parts, 'items' or 'components'). For example: climate, soil, vegetation, mountains and rivers are elements of the natural environment, while examples of elements of the cultural environment are towns, farms, roads and railways. People too are part of geography and are another example of a cultural element.

Fig. 5.1 Interaction is all around us

ISBN: 9780170215701

- In cultural (human) geography two way linkages between places - for example trade between New Zealand and China, and the movement of people between New Zealand and Australia (as migrants or tourists) are examples of interactions.
- In physical (natural) geography the way the sea affects climate and in turn the way climate affects the sea is an example of interaction. How vegetation affects the soil and how the soil affects vegetation is another interaction example.
- People – environment interaction also exists: people causing climate change and climate influencing where and how people live, is an example of people – environment interaction.

Fig. 5.2 Interaction across the globe

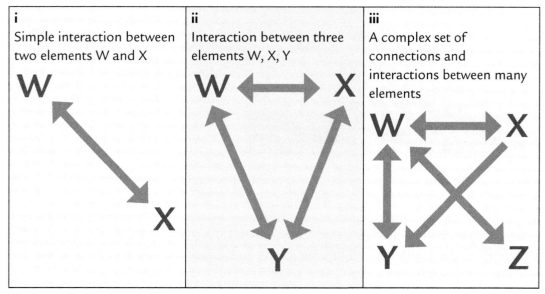

Fig. 5.3 Three examples of interaction

ISBN: 9780170215701

Fig. 5.4 Interaction at the coast

When waves hit the land, the pressure and force of the water can cause the land to erode (wear away). The speed of erosion is much greater when the waves hit the land carrying sand, pebbles and small rocks. Waves are affecting the land – they cause erosion. The land can also influence the waves, for example when a wave hits a cliff some water is reflected back from the cliff into the sea. This reflection interferes with the next incoming wave and reduces the impact of it when it strikes the cliff. Also, as the cliffs erode some of the eroded cliff rocks fall into the sea and can be picked up by the incoming waves. These rocks become 'ammunition' in the wave as it strikes the cliff. This increases the speed of the erosion.

ISBN: 9780170215701

1 a In Figure 5.1 what interaction is taking place in the classroom? Why is this type of interaction not the type that would be studied in geography?

   b Make a rough but large half page copy of Figure 5.2. The arrows show movements, connections, flows and interactions. Label 4 of these arrows with examples of the things that could be moving between the places. For example one arrow could be labelled 'PEOPLE TRAVELLING ON VACATION' and another 'LIVE SPORTS EVENT BROADCAST'.

   c Label these places on your globe:

Atlantic Ocean	Indian Ocean	Mediterranean Sea
Africa	Europe	Middle East
	Asia	

   d Choose the arrow from your globe diagram that could be showing an example of interaction. Describe and explain what this interaction could be.

2 a Make a copy of Figure 5.3 but in each diagram replace the letters with examples of elements. For diagram i. use examples from the cultural world; for diagram ii. use examples from the physical world; for diagram iii. use examples to show people – environment interactions.

   b In Figure 5.3 diagram iii, how is Z connected with X and with Y; how is the X – Y connection different from the X – W connection?

   c Copy and complete the paragraph below. Choose words from the following list to fill in the gaps:

each	changed	landscape	travel
community	around	interaction	glaciers
	moisture	places	

When two things affect _____ other, or when one place affects another place and vice versa it is called interaction. When two _____ have connections then the interaction is 'geographic interaction'. A school and the area _____ a school interact: the area around would be the place students _____ from to attend the school. The school might send cultural groups out into the _____ to perform at primary schools or retirement villages. The two places (the school and the area around the school) are bound together by the _____ that takes place between them.

Two things affecting each other becomes geographic interaction if there is an effect on the _____ as a result of the interaction. Warming climate could melt _____, evaporation from the meltwater could lead to more _____ in the air and higher rainfall. This would be interaction between land and atmosphere. Both the land and atmosphere are _____ as a result of the interaction.

3 Combine the information from the text and three photos on page 64 to draw an annotated diagram to show 'interaction between the sea and land'.

ISBN: 9780170215701

Water connections: people-environment interaction in Egypt and the USA

The studies from Egypt and the USA show interaction between rivers and people in different locations: in the USA between people living in New Orleans and the Mississippi River; in Egypt between people and the River Nile. The Nile and the Mississippi are two of the world's great rivers in terms of their size and their importance to people.

Egypt

An uneven population distribution in any region or country is a normal situation. Egypt presents an extreme example of this pattern: 98% of the population live on just 2% of the land area of the country. It has been this way for thousands of years. The reason for this is that Egypt is a country with very low rainfall. Most of the country is sparsely populated desert. The River Nile provides the lifeline for the country. With headwaters in the wet Ethiopian and East African Highlands to the south of Egypt, the river enters Sudan and then Egypt as a large river. In Egypt the river twists and turns northwards before entering its delta and discharging into the Mediterranean Sea. Most of the population of Egypt live close to the river. This is as true today as it was in the past.

In ancient times, the Nile River flooded predictably each year. During flood the river deposited fine particles called silt on its floodplain (flat low lying land beside a river). This silt provided a fertile soil that was good for growing crops. Most people lived on and around the floodplain area. The river was a matter of life and death in ancient Egypt – if rains failed in the source areas and no floods took place, famine occurred. The calendar of ancient Egypt was based around the annual changes in the Nile. The calendar was divided into three seasons called Akhet, Peret and Shemu. Each season was four months long. Akhet means 'innundation': the time of flooding. Peret was the growing season after the flood waters had gone away and Shemu the time of harvest when the river was at its lowest. In ancient Egypt flat basins separated by earth banks were constructed in the floodplain area beside the river. At peak flood time water was diverted from the river into these basins.

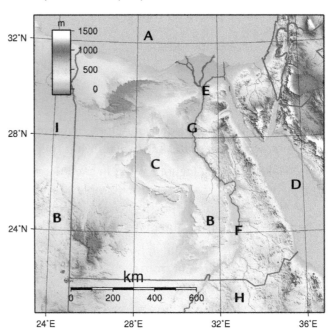

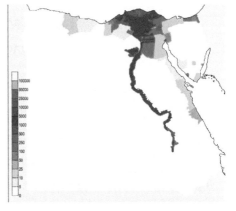

Fig. 5.5 Egypt satellite image and maps showing topographic features and population density (people per sq. km)

The water would reach depths of 1.5 metres. This flooding gave a regular supply of fertile silt to the land and wet soil ready for planting once the flood waters were drained away. In addition to the use of basin irrigation, simple water lifting devices, buckets attached to a water wheel and screwpumps were used to increase the irrigated farming area.

In modern Egypt, the Nile valley remains the place where most Egyptians live. Even though 50% of the population live in cities, these are located along the Nile valley and rely on the river water for their survival. Farming in modern Egypt continues to depend on water from the Nile. Dams now help control the flow of the river and provide a more reliable and controlled source of water. The Aswan High Dam in southern Egypt, the largest of these dams was completed in 1970. This followed the construction of other earlier and smaller dams across the river. As well as increasing the area under irrigation these dams have improved boat navigation along the river, producing a more even water flow during the year, and produce hydro electric power. The loss of regular flooding of farmlands has caused problems however. Farmers must now use large amounts of fertiliser as a replacement for the regular 'natural' input of fertile silt.

The River Nile (part of the natural environment) influences farming and population distribution in Egypt, while at the same time the use of 'artificial irrigation' and dam building shows people having an influence on the natural environment. The river and people interact.

Fig. 5.6 Cities like Cairo and towns, villages and farming in Egypt are concentrated along on the Nile. Desert with low population density lies beyond the river floodplain and delta

Fig. 5.7 In Ancient Egypt, the Archimedean screwpump was used to lift water from the Nile to irrigate farmland

Fig. 5.8 Modern irrigation channel taking water from the Nile to irrigate the floodplain

ISBN: 9780170215701

A Direction of river flow
B River Nile
C Aswan High Dam
D Lake Nasser

Fig. 5.9 An aerial view of the High Aswan Dam

Hurricane Katrina

August 2005: a huge hurricane had been tracked for days as it made its way from the Atlantic Ocean, and into the Gulf of Mexico. This was Hurricane Katrina, a massive storm which was to bring powerful winds, huge waves, torrential rain and devastating flooding in and around New Orleans and across much of the Gulf Coast in the southern United States. It was one of the deadliest and costliest hurricanes ever recorded. More than a million people were forced from their homes, nearly 2000 lives were lost and massive property destruction occurred. People had their lives and futures shattered. Nature had an impact on people that they could do little about.

- *National Geographic* magazine had described New Orleans as 'a disaster waiting to happen'. New Orleans has been flooded many times in its 300 year history. The location of the city at the mouth of the great Mississippi-Missouri River system makes it a great sea and river port. The location is also a problem. The city is built in a bowl (Fig. 5.10). About 50% of the urban area is below sea level, some as much as five metres below. Originally most of the city was built on higher land close to the river, lake and sea shore. As demand for building land grew in the 20th century, engineers drained swamplands. This drainage led to subsidence. Subsidence means land sinking or settling to a lower level. Stopbanks (levees) were built along the river and lake edges as a protection against flooding (Fig. 5.10). The urban area of New Orleans expanded onto lower and lower land. If nature were to prove too strong for the levees the consequences for people and property would be disastrous. Katrina was this disaster. When it struck, the attempts made by people to control nature proved feeble.

- Another factor that made Katrina such a disaster was the loss of wetlands and barrier islands along the Gulf coast. This was an unintended

ISBN: 9780170215701

consequence of people building upstream dams on the Mississippi, which reduced sediment in the river (the sediment was trapped behind the dams) and the levees around New Orleans, diverting river flow and sediment deposition further out into the Gulf of Mexico and away from the coast. Without regular sediment supply, the coastal wetlands and barrier islands had disappeared at a rapid rate between 1950 and 2005. These wetlands and barrier islands provided the city with a natural defence from ocean waves and storm surges.

- Flooding, rather than winds, from Hurricane Katrina caused the greatest problems both short term and long term for New Orleans. Two features of Katrina made the flooding so bad. The air pressure at the centre of the hurricane was a record low (902 hPa) and caused the sea to bulge upwards. This upward bulge is called a storm surge and was nine metres high. This, plus the huge waves driven onshore by the hurricane force winds, overtopped the coastline and coastal defences. Onshore around the city of New Orleans and along the banks of the Mississippi River the levee banks were overtopped and broken through by the rising river and sea waters. The land on which New Orleans was built was swamped. Once flooded the low lying nature of the city added to the problems. When the rains stopped and rivers went down the city was left with deep flood water that had nowhere to go. With land below the sea and river level, the water would not naturally drain away. While pumping stations took time to drain the land the flood water became a stinking and polluted open sewer.

- There is no lack of ideas to protect New Orleans from future flooding. The biggest hurdle is deciding what plans to implement and who is going to pay for them. Rebuilding of the city has taken place and broken flood defences have been repaired and strengthened. Pumping stations have been repaired and added to. Restoration of the natural wetlands and barrier islands offers a longer term natural solution. More controversial is the idea of accepting the geographic reality that converting the lowest lying areas into parks is the best solution – to sacrifice some parts of the city to save the rest. Such a suggestion is controversial because most of the low areas in New Orleans are currently home to poor African Americans who cannot afford to relocate or simply don't want to leave.

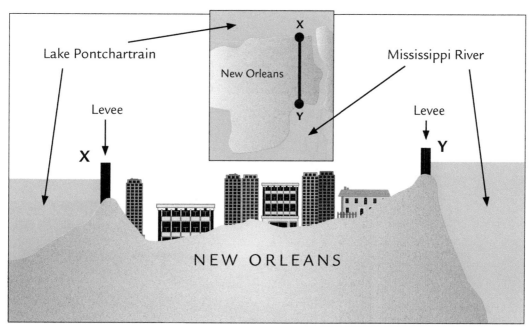

Fig. 5.10 New Orleans: a disaster waiting to happen

ISBN: 9780170215701

Fig. 5.11 Flooding across New Orleans: a city underwater. The Superdome in the centre of the photo, although damaged by Katrina, was used as an evacuation centre.

Fig. 5.12 Homes and businesses destroyed, infrastructure ruined by strong winds and floodwater

Fig. 5.13 Recovering from the devastation brought by Katrina: levees (stopbanks) beside the river are being repaired and strengthened

ISBN: 9780170215701

ACTIVITY

1 a Make a copy of the map of Egypt (Fig. 5.5). Use an atlas to name the features A to I on your map. Choose from these names:

Aswan Dam	River Nile	Red Sea
Libya	Sahara Desert	Cairo
Mediterranean Sea	Egypt	Sudan

b Copy and find the answers (names) to these 12 factual questions from a careful study of this chapter:

i The name of tiny (fine) particles of soil.

ii Name of the land area that a river floods.

iii A word that means 'every year'.

iv Name of a country through which the Nile River flows.

v Large city located beside the Nile River.

vi Artificial watering of crops.

vii Used to lift water up from the river to put on the land.

viii Name of a dam built across the Nile River.

ix City flooded by Hurricane Katrina.

x Month and year that Hurricane Katrina struck.

xi Another name for a river stopbank.

xii Name of the state in which New Orleans is located.

2 a Egypt and the Nile: draw an annotated sketch of one of the Figure 5.6 photographs.

b Refer to Figures 5.7 and 5.8. Explain how irrigation works.

c How would a dam like the Aswan Dam (Figure 5.9) help control river flooding?

3 a Refer to the text and Figure 5.10. What made New Orleans 'a disaster waiting to happen?'

b Hurricane Katrina: Figures 5.11 and 5.12 show ways Katrina impacted on the city and people of New Orleans. Write a description of what happened.

c Figure 5.13 shows one way people are trying to control the river. What does the photo show is taking place?

ISBN: 9780170215701

Inside and outside cities: connections and interactions

Cities and places beyond the city are linked and connected in many ways. Movements take place between them all of the time to the point where they depend on one another – they are interdependent places.

The history of cities can be seen as the history of interaction between cities and the areas that surround them. Today these areas are global rather than local. Cities rely on the areas around them for survival. Food and water from surrounding areas has always been carried into the city: city populations depended on these supplies. In turn, items manufactured in the city like farm tools and handicrafts like carpets were provided for people living in rural areas that surrounded the cities. Today these exchanges involve sales and money, in the past the goods were traded and swapped.

In medieval times the city provided a safe place for local rural people in times of war, where city walls gave protection from invading forces. The industrial revolution of the 18th and 19th century in Europe and later North America and Japan involved raw materials being transported from distant countries to supply the needs of factories producing steel, textiles and food products in new cities located in industrial areas. Many of the manufactured products were then exported for sale across the globe as well as being consumed locally. Today the city offers different attractions like employment for people who commute daily to work in city centre office and shop based jobs.

In mining areas in and around cities like Johannesburg in South Africa, a circular form of migration exists between the city and surrounding areas including neighbouring countries like Botswana. Males leave their families and rural homes and move to work in the mines. Money is sent home on a regular basis, providing an immediate city-rural area linkage. After working in the mines for 5-10 years the miners return to their rural villages and families having between time only had short visits home for vacation.

Cities today offer services that are very much global in their spread. Universities like Cambridge in England and Harvard in the USA enrol students from all over the world. Sports events like the Olympic Games and Soccer World Cup are city based, have teams competing from every continent and attract spectators from every country. At the same time images of these events are transmitted live and to a global audience via TV and the Internet from broadcasting centres in the city.

Cities function as hubs, providing local connections with the rest of the world and vice versa. In the past the hub relied on road and sea transport. Today airports provide this hub

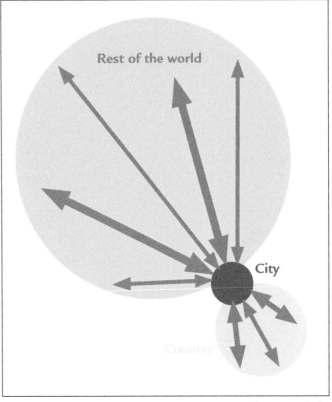

Fig. 5.14 Hub city

ISBN: 9780170215701

function that establishes instant global and local links (Figure 5.14). Airline and shipping offices, hotels, car hire, storage and distribution, post and parcel services and offices of international companies have become features of these hub locations.

Fig. 5.15 Movement in and out of the city

Fig. 5.16 Produced inside, distributed both inside and outside the city

ISBN: 9780170215701

73

Fig. 5.17 People on the move: the daily commute into and out of the city

ISBN: 9780170215701

ACTIVITY

1 a The word hub means 'the centre part of a wheel' and 'the thriving centre of anything and focus of activity'. Make a copy of Figure 5.14 and explain what is meant by 'hub city'. Incorporate the word 'interaction' within your answer.

 b Figure 5.15: make a list or use a star diagram to give examples of five different types of movements shown. One example would be 'by car'.
 Challenge : can you find an example of a movement connected with the natural environment?

2 Figures 5.16 and 5.17: write two generalisations of no more than 25 words describing what these two sets of photos show.

3 Make a large copy of this template. In circles A and B draw sketches to represent the city (A) and beyond the city (B).

 Label the two arrows with examples of movements that take place between the two places.

CASE STUDY 12

Rainforests: soil, vegetation, atmosphere

A car engine consists of a set of parts fed by fuel, oxygen, water and oil. The engine works when the parts are correctly connected. The parts of the engine are interrelated and interaction between the parts provides the power to drive the car. In the natural environment there are lots of similar interactions taking place. Forests for example can be seen as natural green machines that are full of connected and interacting parts (elements).

In a tropical rainforest the vegetation and animals are the most obvious elements to be seen. But below and above the forest are elements that are fed by and feed the forest. Without the soil and climate (rainfall, heat and sunlight) the forest could not survive. In turn, the soil and climate rely on the forest. The three things – soil, vegetation, climate – function as one interacting system. The tropical rainforest environment is a natural environment full of interacting elements.

Fig. 5.18 The rainforest: vegetation and animal life (red and blue macaws, orangutan, leaf cutter ant)

On the floor of the forest, leaves and branches that fall from the trees build up and quickly rot in the warm and wet climate conditions. Humidity is usually at least 80%. This rotting process is helped by animals like ants and termites that eat the organic material and add excrement to the rotting layer. This rotting surface layer is called litter and can be up

ISBN: 9780170215701

to 20 cm deep. Plants gather nutrients and moisture from this litter layer through their root systems. The vegetation is 'making its own soil' to grow in. The root systems of the plants capture over 95% of the nutrients released by the rotting vegetation.

Fig. 5.19 Leaf litter on the ground surface provides the nutrients and moisture for the trees and plants

Above the forest clouds form. The moisture that makes up the clouds comes from water loss (transpiration) from the plant leaves, and from the evaporation of moisture from rivers and wet ground and plant surfaces. Rainforests are among the wettest places on the planet. Rain falls on most days, often in the form of huge afternoon thunderstorms. About 50% of the rain that falls onto the forest comes from moisture evaporated and transpired from the vegetation itself.

Fig. 5.20 Evaporation, transpiration, clouds and thunderstorms

ISBN: 9780170215701

Fig. 5.21 The Amazon Rainforest as seen from space

ACTIVITY

1 a Draw three frames in a vertical format. These frames represent different parts of the rainforest environment. Inside each frame illustrate the frame titles :

Thunderclouds

Plants and animals of the forest

Ground surface litter build up

2 a Explain how these three parts of the rainforest environment are connected.

b Draw arrows between the frames to show these connections and interactions.

3 How does the rainforest environment illustrate the ideas of interaction and interdependence?

ISBN: 9780170215701

patterns

Rectangular farming field pattern and the braided pattern of the Rakaia River channel just below the Rakaia Gorge. Mount Hutt and the Southern Alps are in the background.

ISBN: 9780170215701

GEOGRAPHY DICTIONARY

Patterns are distinctive arrangements of things or objects. Patterns often involve repetitions of designs or shapes. Geography studies spatial patterns: the arrangement of features on the earth's surface, and temporal patterns: how characteristics, trends and features change over time in recognisable ways.

Patterns exist in every aspect of our lives. Patterns are all around us. Some are obvious, others less so. Patterns are repeated designs or arrangements of things like shapes and numbers that can be identified. There are patterns in music, dance, architecture, painting, sports, nature, the seasons and our daily routines. The school day has an arrangement of lesson times, intervals, form time and assembly time. This is a pattern that repeats over time – it may be the same pattern every day or be a weekly cycle.

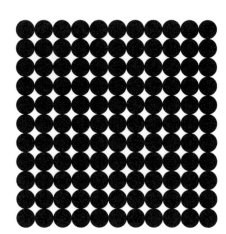
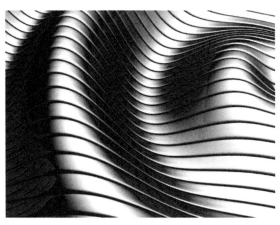

Fig. 6.1 Visual patterns are all around us

Geography has an interest in patterns that exist in present day landscapes, especially patterns to do with location and distribution. Patterns connected with the use of space (land area) are called spatial patterns. Geography also studies how things have changed over time (temporal patterns). Geography identifies, describes and explains patterns that exist in both the natural and cultural world. Geography studies patterns at different scales: local, regional and global.

ISBN: 9780170215701

Fig. 6.2 An urban landscape

Patterns reflect the past and provide a window into the future. The work of nature and the actions of people have created present day landscapes. This has happened over time, slowly in some places and rapidly in others. Looking at the trends – how the patterns have developed over time – provides a way of predicting and projecting into the future.

Fig. 6.3 A rural landscape

ISBN: 9780170215701

1 **a** There are lots of patterns in Figure 6.1. List and describe the patterns you can see in each image. There is no 'one correct answer' and you may see more than one pattern in each image.

   b Use Paragraph A to describe the pattern of your school day or school week.

   c Draw a simple version of the urban landscape in Figure 6.2. Keep the frame size the same as the image. Highlight and label on your drawing the different patterns that can be seen.

   d Describe patterns visible in the rural landscape photo (Figure 6.3).

   e Are Figures 6.2 and 6.3 examples of local, regional or global scale patterns? Give a reason for your answer.

2 **a** Study the spatial patterns in Table 6.1 (below). Name the type of patterns that best describes the shapes shown in Figures 6.1, 6.2 and 6.3. Justify the choices you make.

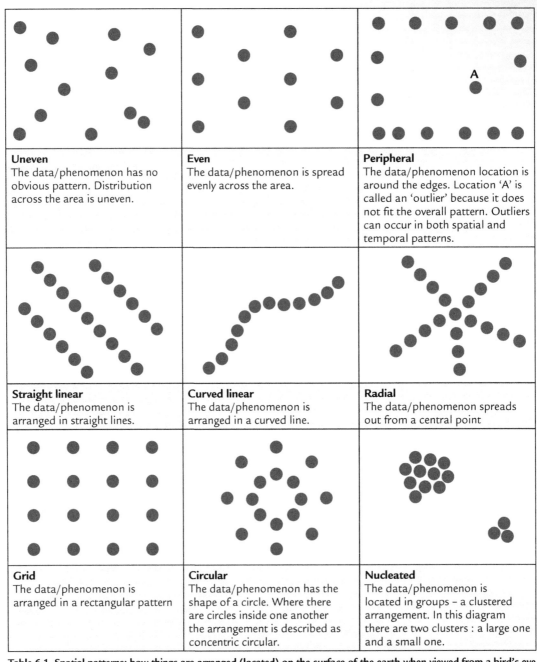

Table 6.1 **Spatial patterns: how things are arranged (located) on the surface of the earth when viewed from a bird's eye (vertical) view**

ISBN: 9780170215701

b Use Paragraph B to explain the difference between spatial and temporal patterns.

c How might the scene in Figure 6.3 change as the seasons change?

3 a Paragraph C reads *'Patterns reflect the past and provide a window into the future'*. How can this be? Reflect on the meaning of this statement, then explain and amplify the meaning of it. How could knowing the past and the present help us to predict future patterns in an area? (Hint: consider how present day landscape patterns of a particular named area have been influenced by the past.)

b Tranform the ideas in Paragraph C into a visual or diagram. A flow diagram would be a suitable type of diagram with past, present and future frames being included.

4 a Use the bold text from Tables 6.1 and 6.2 (below) to complete the 'Patterns' word mix that follows. One example has been done to get you started.

							P							
							A							
							T							
	N	U	C	L	E	A	T	E	D					
							E							
							R							
							N							
							S							

b For each word you include, draw a diagram that shows the pattern.

c Write a single sentence definition for each of these pattern terms: clustered, concentric, cyclic, distribution, equilibrium, fluctuating, linear, peripheral

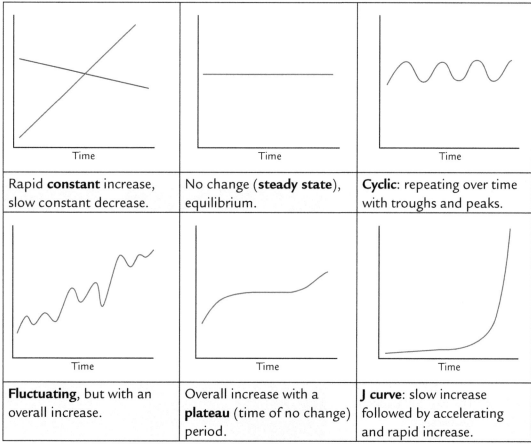

Table 6.2 Temporal patterns: How things change over a period of time

ISBN: 9780170215701

5 Which spatial or temporal pattern best describes each of the following geographic phenomenon? Justify each selection you make with real world examples.

   a Temperature change with increasing altitude.

   b Earthquake activity along plate boundaries.

   c Aircraft routes into and out of a large international airport.

   d Monthly temperatures and rainfall totals for a location over a ten year period.

   e Location of major sports stadia within a city hosting the Olympic Games.

6 **a** Identify and describe the spatial patterns shown in Figures 6.4 and 6.5.

   b Draw a sketch of each location. Add the labels given in the title.

   c Write a paragraph comparing and contrasting the patterns shown in the two photos.

Fig. 6.4 Rakaia River, Canterbury Plains and the Southern Alps

Fig. 6.5 New York City: Manhattan, Central Park and the Hudson River

7 **a** Describe each of these patterns visible within Figure 6.6.

   i Pitch markings.

   ii Shape of the pitch.

   iii Location of the crowd in relation to the pitch.

   b Imagine you were looking at the stadium from above (a bird's eye view). Draw a map or sketch of what you would see. Include labels and/or a key.

   c Suppose there are two teams playing, and a scrum forms on the halfway line close to the touchline by the entrance tunnel. Show on your map or sketch:

     i The scrum as a clustered pattern.

     ii The two backlines making linear patterns.

     iii The two full backs as outliers.

Fig. 6.6 Stadium patterns: Waikato Stadium, Hamilton

ISBN: 9780170215701

Taranaki: examining spatial patterns

The landforms of Taranaki/Egmont National Park shown on the topographic map (Figure 6.7), aerial photograph (Figure 6.8) and the photo of (Figure 6.9) are all natural.

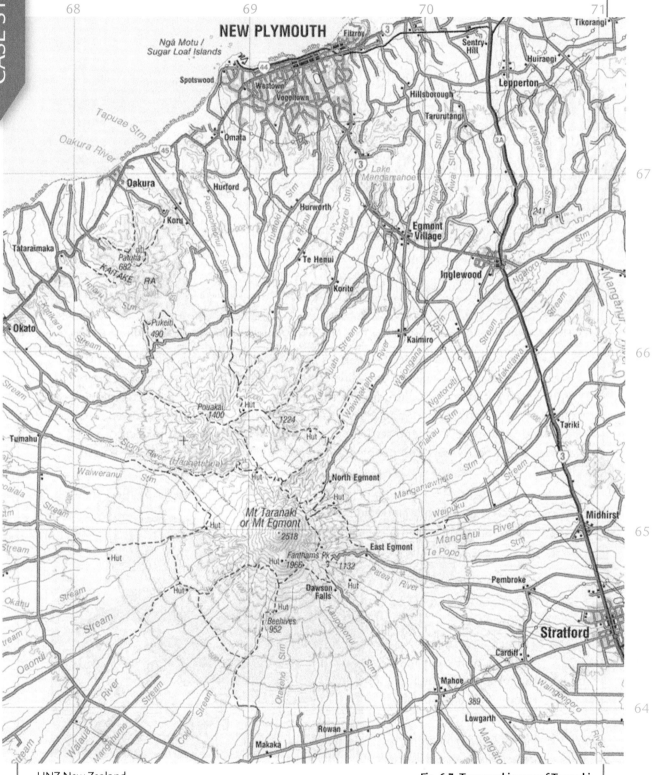

LINZ New Zealand

Fig. 6.7 Topographic map of Taranaki

KILOMETRES

ISBN: 9780170215701

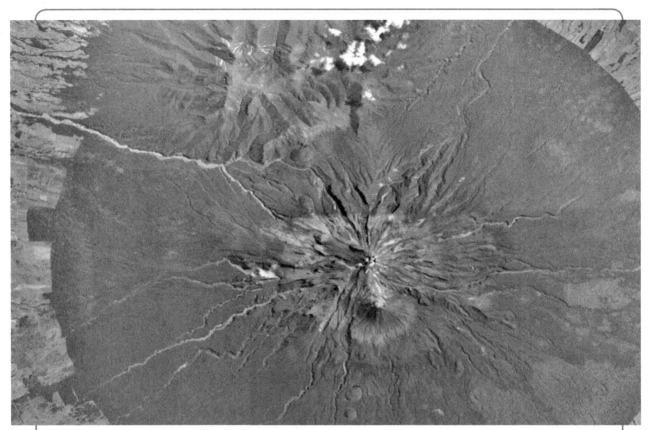

Fig. 6.8 Aerial photograph of Egmont National Park, Mt Egmont/Taranaki, Fanthams Peak, Stony River and Pouakai Range

These landforms are the result of volcanic activity, weathering, mass movement and river erosion. The upper tree line and snowline on the mountain are also natural, reflecting reduced temperatures with altitude. The sharp lower tree boundary shows the influence of people. The national park boundary sets the limit of forest clearance allowed to create farmland. Transport, farm and settlement patterns are the results of the decisions and actions of people. Nature and people have created the present day landscape patterns.

ACTIVITY

1 Refer back to Table 6.1 to explain the type of pattern that best describe these features of Taranaki and Egmont National Park.

 a The drainage (river) pattern.

 b The road pattern.

 c The vegetation pattern.

2 Draw annotated sketches of the following.

 a The aerial photo (Figure 6.8). Refer to the topographic map (Figure 6.7) to help with this task.

 b Figure 6.9, highlighting the patterns and other features it shows.

Fig. 6.9 The landscape of Taranaki

3 a How does altitude influence the pattern of rivers, transport and vegetation within Taranaki/Egmont National Park?

 b Describe the settlement pattern shown on the topographic map (Figure 6.7).

ISBN: 9780170215701

World population growth since 1700: examining a temporal pattern

The world's population is now around seven billion. There has been more growth in population in the last fifty years than the previous two million years. Change in the global population size is due to just one factor: the balance between births and deaths. On a global scale for most of human history the number of births has been greater than the number of deaths. This has resulted in a growth in the total number of people living on the planet. Before the year 1800 global population growth was slow. Even between 1800 and 1900 the rate of growth, although increasing, was slow when compared to the phenomenal growth that took place during the 20th century.

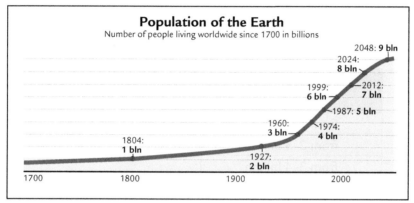

Population of the Earth
Number of people living worldwide since 1700 in billions

2048: **9 bln**
2024: **8 bln**
1999: **6 bln**
2012: **7 bln**
1987: **5 bln**
1960: **3 bln**
1974: **4 bln**
1804: **1 bln**
1927: **2 bln**

1700   1800   1900   2000

Fig. 6.10 Population growth on Earth from 1700 including projections to 2048

In the last 200 years, advances in medicine and other medical technologies have led to a lowering of the death rate and a rapid increase in life expectancy in wealthy nations. These nations as a group became labelled as the developed world. In most of these nations today, especially those in Europe, population growth has slowed due to a declining birth rate. In some nations like Ukraine, Poland, Italy, Germany and Japan there is an actual decline in total population because the number of children being born is below replacement level: deaths now exceed births. Death rates in these nations have also increased because of the ageing of the population, so there are more elderly people. In this group, death rates are 'naturally' high due to age related factors like heart attacks and cancer.

The continuing rapid increase in world population since 1950 has been due to events in countries within Latin America, Africa, the Middle East and parts of Asia. The spread of modern medicine, with more qualified health professionals, modern clinics and hospitals has allowed for the control of infectious diseases and the prolonging of life. Babies survive their early years and people are living longer. Today, rapid population growth in a lot of Latin America, the Middle East and Africa is due the very high birth rates of 20–40 years ago, which have resulted in large numbers of people currently being in their reproductive years. Although birth rates are declining, there are more women to give birth, hence population growth continues. This phenomenon is called population momentum. This momentum will work its way out of the population as the large number in the reproductive ages move into later middle age and old age.

On the global scale, decreased population growth rates are becoming apparent even though total numbers continue to increase. Demographers (population experts) predict that by the end of the 21st century the world's population will stabilise at nine to ten billion people. In the meantime, in absolute numbers it is still growing faster than ever before, by about 230 000 people a day.

ISBN: 9780170215701

ACTIVITY

1 a Refer to the population growth graph (Figure 6.10) and back to Table 6.2
      (temporal patterns) to describe the pattern of world population growth:

      i between 1700 and 1900.

      ii between 1900 and 2050.

   b What does the word 'temporal' mean?

   c Which temporal pattern graph best matches the overall change shown in world
      population between 1700 and 2050?

   d Write the following numbers as numerals. (Follow the example for 1000.)

      One thousand = _1000_

      One hundred thousand = _____

      One million = _____

      One billion = _____

2 a Use the statistics in Figure 6.10 to complete this table.

Growth in world population	Number of years it took or is projected to take
from 1–2 billion	
from 2–3 billion	
from 3–4 billion	
from 4–5 billion	
from 5–6 billion	
from 6–7 billion	
from 8–9 billion	

   b Refer to the text on page 87: calculate how many millions of people the world's
      population is increasing by each year at the present time.

3 a Explain why the rate of global population growth has changed over the last 300
      years (from 1700 to the present day).

   b How is the growth in global population projected to change between now and the
      end of the 21st century?

   c What is the basis for making such projections?

   d What things could cause the projections to be wide of the mark as time proceeds?

ISBN: 9780170215701

CASE STUDY 15

Visible light at night: spatial patterns

These two world at night images (Figures 6.11 and 6.12) were created from satellite data. The images show an uneven pattern of night light across the world and within individual countries like the USA. The brightest areas of Earth are the most urbanised and technologically advanced, but they may not necessarily be the most populated. Cities tend to grow along coastlines and transportation networks, so even without the underlying map, the outlines of many continents would still be visible. In Russia, the Trans-Siberian railroad is a thin line stretching from Moscow through the centre of Asia to Vladivostok. The Nile River, from the Aswan Dam to the Mediterranean Sea, is another bright thread through an otherwise dark region.

More than 100 years after the invention of the electric light, some regions remain thinly populated and unlit. Antarctica is entirely dark. The interior rainforests of Africa and South America are mostly dark, but lights are beginning to appear there. Deserts in Africa, Arabia, Australia, Mongolia and the US are poorly lit as well (except along the coast), as are the boreal forests of Canada and Russia and the great mountains of the Himalaya.

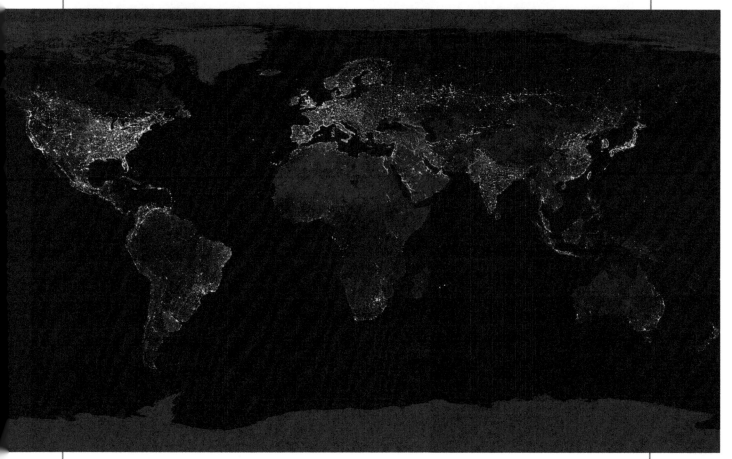

Fig. 6.11 The whole planet at night: a small scale view

ISBN: 9780170215701

The US interstate highway system appears as a lattice connecting the brighter dots of city centres.

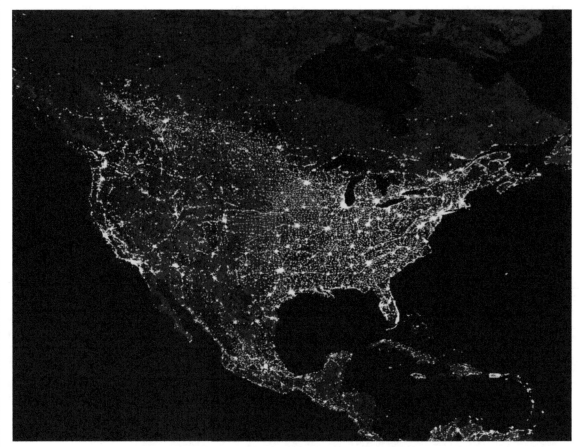

Fig. 6.12 North and Central America at night: a closer, larger scale view

ACTIVITY

1 Refer back to Table 6.1 to describe the pattern of global night light shown in Figures 6.11 and 6.12.

2 Use an atlas to help complete these two tasks.

a Identify and name three areas in Figure 6.11 with a lot of light and three areas with little light.

b Based on a study of Figure 6.12 draw an outline map of North and Central America and then shade and name specific locations like cities, lakes or mountain areas that show:

i A lot of visible night light.

ii Little visible night light.

3 a Refer to Figures 6.13, 6.14 and 6.15, which show the cities of Auckland, New York and Hong Kong at night. Imagine that you are flying into one of these big cities at night. Looking down from above you would see patterns of light and dark. Describe what a typical pattern might look like. What would create light? Where would it be dark? Some features and activities to think about: airports and seaports, railway stations, roads, motorways and freeways, housing suburbs, sports grounds, river and sea areas, entertainment events, parks and industrial areas.

b Looking at the text and images in this case study, explain why the amount of night light across the globe displays spatial variation.

Fig. 6.13 Auckland city seen at night from Sky Tower

Fig. 6.14 New York seen at night: Brooklyn Bridge and Manhattan

Fig. 6.15 Hong Kong airport at night

ISBN: 9780170215701

7

Perspectives

Perception
Kaitiakitanga
Tino Rangatiratanga

*Oil and gas prospecting and drilling brings
debate and polarised views.*

ISBN: 9780170215701

GEOGRAPHY DICTIONARY

A **perspective** is a way of viewing the world. Perspectives influence the way people view and interpret environments. Perspectives can influence how people interact with environments and the decisions and responses that they make. Perspectives may be influenced by culture, environment, social systems, technology, economic and political ideology.

Perception is the view/opinion we have about things and how we see things. Our perception (view/opinion) is influenced by the perspective we come from. Mountain bike riders, conservationists and mining companies are likely to have a different view about activities that should be allowed in a national park. They come from different perspectives and therefore would have different views about acceptable park uses.

It's all about perspective

A friend had just completed the marathon (42 km) in a good time of three hours and thirty-five minutes. While standing in the finish area, he found himself chatting with the winner, who had run the race in two hours and five minutes. A woman approached the two of them. 'I want to congratulate you,' she said to the winner. 'And tell you how much I admire you. I just finished my first marathon in a time of just under six hours, and here you are running the race in close to two hours. You are amazing.'

The winner paused and thought for a moment about what the woman had told him. 'Actually, ma'am,' he replied, 'you are the amazing one. I can't imagine running for six hours.' The top marathon runner and the first timer both viewed one another's performances with awe. Time meant different things to two runners – they viewed what was an amazing time from their different marathon experiences.

ISBN: 9780170215701

We may see and read about the same things and same events, but the way we interpret and respond to them is not always the same. Why? Because we have different perspectives. How we see the world depends on our perspective.

The perspective that people have is influenced by their values, previous experiences, culture and identity. Perspective is influenced by who people are and where they come from.

From the angle of geography this means people will interpret and interact with the world in different ways depending on their perspective. Perspectives may be individual (this can be called a point of view), or a view shared by many. A view or standpoint shared and recognised by many people is often called a 'world view'.

Perspectives that are relevant to geography

Economic perspective	Social perspective	Gender perspective	Communist perspective
Green (environmental) perspective	Maori perspective	Multi-cultural perspective	Indigenous people's perspective
Age related perspective	Capitalist perspective	Religious perspective	Local perspective
Scientific perspective	Futures perspective	Global perspective	Mono-cultural perspective

Table 7.1

There is no right or wrong perspective

Perspective is about a point of view. Everyone has their own view, and there is nothing that draws a line that says 'this view is right' and 'that view is wrong'. What seems to be correct to some may not be so for those with a different perspective. It all depends upon the perspective of the individual. You could say, 'You have your own way, I'll have my own way. As for the right way, the correct way and the only way, it does not exist.'

Fig. 7.1 Different views of a city: the things that we see and how we react to them depends on our perspective, and not always how things are in themselves

ISBN: 9780170215701

1　a　In the marathon story what was the time of the winning runner? What was the time of the first time competitor?

　　b　The winner and first timer in the marathon both admired and thought amazing the time of the other person. Why was this?

　　c　The title to the marathon story says 'It's all about perspective'. Explain what this means.

2　What do the words in Table 7.1 mean? Copy and complete the table below by selecting a perspective that would match the focus of interest.

Perspective	Focus of interest
	Creation of profit, finances, jobs and wealth
	Conservation and environment well being is all important
	Seeing things from the viewpoint of an elderly retired person
	Taking a broad view about the whole world rather than about local issues and local impacts
	Concerned about culture and respecting links to the past
	Having a focus on whether things can be proven by facts and experiments

3　Figure 7.1 shows a city viewed from above (a vertical perspective) and the same city viewed from the ground (a horizontal or ground level perspective). Imagine an incident has taken place in the city. You and a colleague are reporting on events for the local radio station. You are in the police helicopter above the incident scene. Your colleague is with the ground level police team.

　　i　Decide what the incident might be, such as a robbery, car crash, kidnapping, hold-up or a domestic.

　　ii　Prepare a the transcript (text) for two 30-second-long live radio reports. The first should be 'your report' to be broadcast from the helicopter, the other is for your colleague to be broadcast from street level.

　　iii　Each report needs to describe the city (features and nature of the city from the vertical and ground level perspectives), and an outline of what is happening (in terms of the incident). Be imaginative but realistic in your account and description.

ISBN: 9780170215701

Oil prospecting and drilling

Oil and gas exploration and production takes place all over the world. Some production like in the Middle East and central Asia takes place mostly on land. In other parts of the world such as the North Sea in Europe, exploration and production takes place offshore or out at sea (Figure 7.2). In the southern US and Gulf of Mexico there is a mixture of onshore and offshore production (Figure 7.3).

In the Taranaki region of New Zealand both onshore and offshore wells have been drilled and are in operation. Oil, gas and the products made from them are essential in our world today, however, exploring for oil can be controversial and not always welcomed by local communities. Nevertheless, oil and gas are needed as fuel for transportation and heating, and as a raw material used in the manufacture of a wide range of products like plastics, textiles, paints, chemicals and medicines.

Fig. 7.3 Oil drilling rigs and supply vessels operating near shore

Fig. 7.2 Oil wells are often drilled in remote offshore locations accessible only by sea or air

Different perspectives

From a government and local perspective the thinking might be that development of oil and gas fields is positive and beneficial. After all it would bring income from taxes paid by the oil companies, as well as create employment onshore (the oil and gas would need to be transported and unloaded) and on the rigs themselves. Many such jobs would be well paid. The local economy could profit by supplying equipment and services to the oil rigs. Local towns and local people could therefore benefit.

Not everyone might be so positive, however, and environment groups often express concerns about possible pollution if there were oil spills, or the dangers of fires on oil rigs. There are regular reports from all around the world of environmental disasters when things go wrong (like in the Gulf of Mexico, Alaska and Nigeria). The grounding of the container ship *Rena* off the Bay of Plenty coast in 2011 brought home the damage oil spills can cause. Thick black oil covering beaches, fouling rivers and clogging bird wings is not a good look no matter how much clean up takes place. Worse still, all life within the area is threatened if water and farmland becomes covered in oil or contaminated by other chemicals. Local economies relying on tourism also suffer when such disasters occur.

ISBN: 9780170215701

Kaitiakitanga: this concept emphasises the importance of conservation and caring for the environment. It refers to Maori environmental management systems developed to protect and enhance the mauri of taonga and ensure the sustainable management of natural resources. It is about the sustainable use, management and control of natural and physical resources that are carried out for the mutual benefit of people and resources.

Tino Rangatiratanga: is the right to self determination and the authority to protect and preserve and make decisions for yourself. Kaitiakitanga can only be practiced through the exercise of tino rangatiratanga. Rights, responsibilities and obligations involving the use, management and control of the land and other resources are essential components of tino rangatiratanga.

Big oil and gas exploration permit awarded for the Raukumara Basin

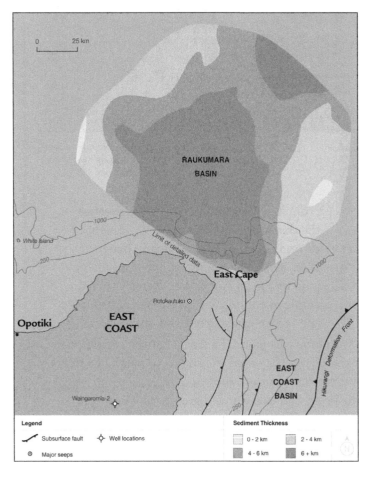

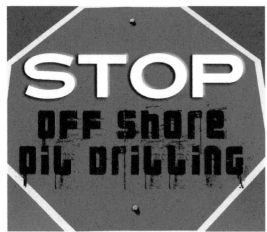

ISBN: 9780170215701

JUNE, 2010:

The government has awarded New Zealand's first petroleum exploration permit over the Raukumara Basin off the North Island's East Coast to Petrobras, a Brazilian oil and gas company.

"Petrobras is an international giant and a world leader in development of offshore drilling technology and production. This is an exciting step into areas of New Zealand until now unexplored," the energy minister said. "If the exploration is successful and development goes ahead it will be an important part of a better future for all New Zealanders - bringing more jobs, more tax and royalty income, and most importantly, creating opportunities for long term regional growth".

Government MP Hekia Parata (grew up in Ruatoria near Gisborne, of Ngati Porou and Ngai Tahu descent) explained her support for the project was based on the economic benefits it could bring to the East Coast region. Exploration and development had given Taranaki, NZ's largest oil and gas producing region, a significant boost, she said. In 2009, Taranaki's oil and gas industry employed 5090 workers and contributed $2 billion in GDP. The average salary in the oil and gas industry is about $70,000 per year, compared with $44,000 in the tourism sector.

"I love the East Coast and want to see it get similar benefits that Taranaki has got from oil and gas development," she said. She went on to say that she understood the concern people had over the impact of the petroleum industry on the environment, but said the Government was determined to ensure New Zealand's marine environment was properly protected as petroleum resources were further developed. She added that it was in the interests of Petrobras to operate in an environmentally friendly manner.

Not everyone was so positive about this announcement. Greenpeace and East Coast Maori raised objections and protested.

"The ocean and these coastlines are the New Zealand economy - they are also our way of life and they are a national treasure, too valuable to risk for oil money," said Steve Abel, Greenpeace's climate change campaigner.

April 2011: Police have laid charges against the skipper of a protest boat who was arrested while disrupting an oil survey ship yesterday. Elvis Teddy, the captain of the boat San Pietro, was arrested for breaching the Maritime Transport Act after police boarded his boat yesterday morning amid protests against Brazilian company Petrobras's search for oil.

San Pietro, manned by local iwi, was stationed, along with three other protest boats, in front of the Petrobras survey ship *Orient Explorer* in the Raukumara Basin, off the coast of Gisborne, police said.

San Pietro is owned by East Coast iwi Te Whanau a Apanui and is part of the flotilla including Greenpeace and the Nuclear Free Flotilla, in its third week of opposing deep sea oil drilling.

ISBN: 9780170215701

MARCH 29, 2011:

Letter to the editor: **Protecting sea, land and people**

Aue Hekia (Parata), Te Whanau a Apanui and our other iwi and hapu on the East Coast are saying no to drilling, your own iwi is against drilling, our kuia and koro are against drilling. The haukainga who live and keep your homefires burning are against drilling.

You say there will be plenty of opportunity for public input but what about the iwi and hapu, kuia and koro who are already telling you publicly to stop any drilling and mining?

We don't need a flash degree to know that if anything should go wrong when drilling starts on the East Coast, it could destroy this economy and Maori along with it.

Why? Because Papatuanuku and Ranginui are us and we are them and we have no right to place our Atua in danger.

Yes, the people need jobs but not if it places the very food and water we live on at risk.

We didn't want it, we didn't ask for it so don't give it to us. Your government should be thrown out for making bad decisions.

Petrobras, you are causing havoc on the East Coast. You are not welcome here.

ACTIVITY

1 a List the different areas of the world referred to where oil and gas drilling takes place.

 b Name four different ways oil and gas are used.

 c The case study states that oil and gas 'are essential in our world today'. Refer back to Table 7.1 on page 95. Which of those perspectives would such a statement be based upon?

2 Refer to the text under 'Different perspectives' (page 97). Draw two star diagrams, one for the 'National and local perspective' and one for the 'Environmental perspective'. On the points of each star put the thinking and feelings associated with oil and gas rig operations.

3 Choose two other perspectives from Table 7.1, preferably one of them a Māori or indigenous person perspective. Describe the thinking and feelings that people coming from these perspectives might have about the oil and gas rig operations.

4 Draw a map or write a paragraph to show or describe the location of the Raukumara Basin.

5 Make a list of positives and negatives associated with the Petrobras exploration and drilling plans.

6 Why do the two sides (government/Petrobras and local iwi/Greenpeace) have such opposite reactions to the awarding of the exploration permit? Refer to Kaitiakitanga as part of your answer.

ISBN: 9780170215701

The polar bear and the wild

A 17-year-old school student on a scientific and adventure trip to the Norwegian Svalbard islands in the high Arctic was mauled and killed by a polar bear in August, 2011. Four other students in the group were also attacked and suffered bite and claw injuries. They were airlifted to hospital and later recovered. The polar bear was shot and killed by one of the group leaders. Left behind are the grieving family of the student who died.

Two reports on the event

A

> ### Student attacked, mauled and killed by polar bear. The group leader was a hero and shot it dead ...
>
> - 17-year-old victim had been camping on remote glacier on a wildlife trip.
> - Bear was punched away by a boy lying next to victim.
> - Days before attack, group was 'delighted' at seeing a polar bear.
> - Four other victims were taken to a nearby hospital.
> - The bear was 'looking for food' in town with history of attacks, say residents.
> - A group leader shot the bear dead with a rifle.

Fig. 7.4

B

Experts say that at this time of the year (the northern summer) the sea ice is too thin for bears so they have to stay on land. With food scarce the bears can become so hungry that they will ignore alarm signals and warning shots and attack camp sites and campers for food. Under local laws the bears are protected and it is prohibited to seek them out and disturb them. Local people said that visitors all too often forget the polar bear is a wild and dangerous animal and that we are guilty of invading their habitat.

Some local experts asked the question of whether people can have things both ways: humans want to see animals in the wild and in their natural habitat, but when animals attack us as invaders into their territory – a natural instinct – we cry foul and take action against the animal which in this case was to shoot it. The death of the student is awful and our sympathies go out to the grieving family. It is worth remembering though that we choose to go to remote habitats and become involved with nature.

These are the habitats of wild animals. So there remains another tragic outcome. For the polar bear family – the female partner and cubs now have to look after themselves and find food if they are going to survive.

Fig. 7.5

ISBN: 9780170215701

Polar bears: background facts

Polar bears are one of the few wild species which will actively hunt humans.
At 3 metres tall and half a tonne in weight, they are the world's biggest land predators and top of the food chain in the Arctic.
These creatures can smell prey 30 km away, smash through metres of ice in minutes to reach seals and devour 50 kg of meat at a time with their razor-sharp teeth.
They have incredible vision, can run on the ice at 40 kph and are also powerful swimmers capable of crossing 50 km of water at a time, making them extremely difficult to escape.
Although they feed chiefly on marine animals such as seals and young walruses, they are fearless and will stalk any animal when hungry, including humans.
There have been several previous polar bear attacks on humans in Svalbard, the area where the British teenager was killed.
Last summer a polar bear tore a Norwegian camper from his tent and dragged him 50 m across ice and rocks while he was on a kayak expedition in Svalbard. Sebastian Plur Nilssen, 22, suffered cuts to his chest, head and neck, but survived by grabbing a rifle and killing the bear with four shots.
Locals said there have also been attacks on a man from Austria and a girl, who both died.
Liv Rose Flygel, 55, an artist and airport worker from Svalbard said, 'It's not the first time. The problem is when the ice goes the bears lose their way and cannot catch food. People don't really know how dangerous they are. One came down to the sea recently and people were running down to take pictures.'
Polar bears are well adapted for surviving their hostile, barren environment.
Their double layer of fur and four-inch thick layer of fat means they can live in temperatures of minus 50 ˚C.
During the warmer seasons, the bears mate and give birth as they wait for the ice to form, usually in October.
Scientists say there are 22 000 to 27 000 polar bears in the world, 60% of them in Canada. They also live in Alaska, Russia, Greenland and Norway.
Polar bears are now considered a 'threatened' species as the total number of polar bears has fallen to 25 000. However, hunting restrictions have helped the population to stabilise.

Table 7.2

1 a Read the newspaper accounts in Figures 7.4 and 7.5 and then write your own summary of what took place in 40–50 words.

   b Based on Table 7.2 present a short fact file/infographic (a set of important information presented in a visual and eye catching way) about polar bears.

2 a Why do polar bears attack humans?

   b The two accounts in Figures 7.4 and 7.5 take a very different approach to reporting the incident. How do the perspectives of the two writers differ? What perspectives lie behind the two reports?

3 Outline a plan of action to prevent tragic incidents like this one happening again.

ISBN: 9780170215701

CASE STUDY 18

Brand New Zealand and the clean green 100% Pure image: fact or fiction?

A **100% Pure – as true a slogan as it ever was**

New Zealand in the eyes of anyone who has seen the world is notably green, clean and fresh. Tourists attracted by the 100% Pure slogan are unlikely to be disappointed. The 100% Pure and clean and green branding used in marketing changes nothing, but simply reinforces what exists.

We have a favourable climate without the extremes of heat, cold or drought. The climate and weather are variable but tourists marvel at these changes as they travel the country. Tourists come and experience stunning and varied landscapes, and nature at their doorstep – or at least at the doorstep of their campervan. Bathing in hot pools surrounded by snow and ice and seeing glaciers coming down to sea level provide experiences few other countries can offer. Wilderness experiences go hand in hand with adventure tourism like black water rafting and heli-skiing.

Our coastline is long, varied and undeveloped compared with most overseas countries. Most of it remains in a natural state without the blight of coastal high rise and intense subdivision. The country is a paradise for the geographer as well as for the tourist.

Farming works in balance with nature. We rely on farming for our income and farming remains the backbone of the nation. We lead the world in having environmentally friendly agricultural practices that are well regulated by farmers themselves as much as by government. British consumers watch television adverts for NZ butter. A pair of cartoon cows graze on a green hillside reading an upside down map. The voiceover says, 'Only our cows are free to roam all day long. Anchor – the free range butter company'.

Environmental standards are ever improving. We need to use the environment for the benefit of everyone but this is being done in a careful and environmentally friendly way. Native plants and

ISBN: 9780170215701

trees are being replanted, indigenous birds and animals protected and rural land use recognises the importance of environmental stewardship. Farming and conservation work hand in hand.

44% of our land is covered in native vegetation and 32% of the country is conservation land (protected from development).

New Zealand does not have to try very hard to be clean, pure and green. With a low population density, surrounded by oceans and exposed to onshore winds we are guaranteed fresh air, frequent blue skies, bright light, plenty of rain, flourishing vegetation and wide open spaces.

B The myth of a clean, green Aotearoa

From a letter to the editor: 'Many Europeans travelling to NZ ask my advice on where to go surfing, windsurfing, kiting and so on. Imagine my embarrassment when we visit a legendary local surfing spot along Taranaki's surf highway 45. The water is brown with sludge and contains floating livestock carcasses. I am now recovering from a stomach upset probably contracted from surfing at Orewa just north of Auckland. I know I should not go out after heavy rain when storm water and sewage pipes overflow into the sea, but how do you explain to tourists that although the skies are now clear and the sun shines down 'We cannot go swimming until Thursday'.

The Manawatu river has been labelled as one of the most polluted in the western world after tests showed it heavily polluted from farm run-off (fertiliser, urine and excrement), treated town sewage and factory waste. This is not an isolated example: 70% of the streams and rivers running into the Waikato River have tested as unsafe for swimming. Craig Potton, landscape photographer, conservationist, tramper, climber and maker of the 'Rivers' TV series came to a simple conclusion, 'We don't love our rivers enough'. Four of the five rivers in the TV series he found were under serious threat and 'slowly dying', the Waikato and Rangitata from intensive dairy farming and the Clutha and Mokihinui from hydro power developments. Only the Clarence River seemed safe – for the moment at least.

At many river monitoring stations, water quality is getting worse rather than better. Dairy farming is seen as the main culprit. More dairy farms and more intensive dairying is placing huge stress on our waterways.

Reports on the state of Auckland harbours show significant water pollution. Some of the problems are related to streams that drain into the harbours. These are streams that drain urban catchments not rural land. The Puhinui flows through the suburbs of south Auckland and drains into the Manukau Harbour. It carries with it road run-off – petrol, diesel, tyre and brake pad grime. Added to the mix are litter – paper, glass and plastic – that makes its way into the drains. Then there is the run-off from gardens and garden waste, plus illegal factory dumping. The image is far from clean and green, and the smell tells you it is not 100% pure. The Māori saying that 'the ocean starts at the mountain top' seems apt.

Taking the national picture it seems clear that any clean green image we have is based on myth and marketing. Our small national 'environmental footprint' (measuring resource use and waste creation) is based on us having only 4.5 million people. On a per capita basis our footprint is amongst the world leaders. Not a record to be proud of.

ISBN: 9780170215701

ACTIVITY

1   For each side of the argument (A: 100% Pure and B: The myth) draw an annotated sketch of one photo and write a description of the other.

2   Based on the evidence presented, which side (A or B) seems to present the picture of the 'real New Zealand'? Give reasons for your answer.

3   These two very different accounts about the state of the New Zealand environment could be described as being written from two different perspectives. Choose two perspectives from the list in Table 7.1 on page 95 that you think best match the two accounts, and justify your choices.

ISBN: 9780170215701

Processes

8

Systems
Migration
Hekenga

*From orchard to processing factory and
on to the consumer.*

ISBN: 9780170215701

'Process' is a word in common use. Dictionaries supply these definitions:

- A series of actions or changes that bring about a result e.g. the process of obtaining a driver's license.
- A sequence of interdependent and linked procedures that transform inputs into outputs, such as processing milk into cheese, or converting computer data from one form to another. At the end of a process are the results or outcomes. These results can be seen as the 'destination' and the process can be thought of as the 'vehicle' that gets you there.

The geography definition of process reflects these definitions because it emphasises a sequence or series of actions (steps) but adds that these actions need to be related to environments, places and societies.

Summary: A geographic process involves a sequence of actions or events related to people and/or the environment with a spatial connection. Some geographers say, 'to understand processes provides an understanding of the world as it is today.' Processes are part of the geography of our world and help explain features of our world.

Flow diagrams are an effective way of showing the operation of a process

A: Sequence of events with a final outcome

B: Cumulative causation

Fig. 8.1

ISBN: 9780170215701

Written description of a cultural process

The development of a new office and commercial park near the airport meant there were jobs available. Many people, especially the well qualified, moved into the area. New housing areas next to the harbour were developed on the peninsula a 15 minute drive away from the airport. The census showed the area was the fastest growing part of the city. The old main road shops underwent a facelift and many new restaurants opened. Nearby a new fitness and health centre was built. Local school rolls increased. There were plans for a new high school to be built. Traffic congestion became a problem as the new services and facilities in the area became an attraction for many people. The area was on an upward growth spiral.

Natural processes – a photographic study

Fig. 8.2

Wind dragging over the ocean creates waves (Figure 8.2 A). The energy of the wind is transferred to the waves. As the waves get close to shore and enter shallow water they get higher and steeper (B). They crash against the shore with great force (C). If they have picked up sand and stones as they get close to shore this material as well as the force of the water hits the coastal cliffs (D). Coastal cliffs get worn back by the wave action. With time property on the cliff edge comes under threat of collapse into the sea.

ISBN: 9780170215701

Process is an important and wide ranging concept in geography. When giving and studying examples of geographic processes, the processes selected can themselves be examples of geographic concepts. For example, in Figure 8.2 'wave erosion' is highlighted. Erosion is a geographic concept in its own right. In case study 19, migration (hekenga) is used to show the operation of a (cultural) process. Migration is also an important geographic concept that is widely studied in geography. Process, erosion and migration are all examples of concepts that are important to this subject.

Some concepts used in geography are very broad, such as patterns, change and processes, while others, like migration, globalisation and the water cycle are much more specific and focused.

ACTIVITY

1 a Use a flow diagram similar to Figure 8.1 A to show the process involved in obtaining a driving licence.

b Study the flow of apples illustrated in Figure 8.3.

i Suggest what events would be taking place at stages A, B, D and F. (Hint: Event A should be taking place on an orchard.)

ii Describe what is happening at stages C and E.

iii What process is being shown in this flow diagram?

iv How could this process be related to places, the environment and people?

Fig. 8.3

2 a Refer to Figure 8.2 and the text underneath it. Describe what is happening in each picture. What natural process is taking place?

b Over time what is likely to happen to the cliff shown on Figure 8.2 C?

c Make an annotated sketch of Figure 8.2 D.

d What threat from the sea is there for houses shown in this photo?

3 a Read the written description of a cultural process. Make a large copy of Figure 8.1B and add information to each arrow using details from the description. Title the completed diagram 'An upward urban growth cycle and the process of cumulative causation'.

ISBN: 9780170215701

b   Is this a 'geographic process'? Based on the information on pages 108–110, complete the following to show three things that could be part of a test to judge whether something is an example of a geographic process.

   i   There needs to be _____

   ii There should be _____

   iii There has to be an outcome that _____

Systems

A system is an example of a process but it is also a concept in its own right. A system is a set of connected natural or cultural phenomena (elements, parts or objects) which work together for a purpose, or that form a whole. Geographic studies are intererested in many different systems: for example mountain systems, beach systems and river systems from the physical world; transport systems, manufacturing systems and farming systems from the cultural world.

   Systems are dynamic. This means they are full of movements and are always changing – if there is a change in one part of the the system then further change is likely to take place within the system. Systems are often involve inputs, transfomations and outputs which are linked one to another.

ACTIVITY

1   The water cycle (hydrological cycle) shown in this diagram is an example of a system. Make a large copy of the diagram and on your copy add the labels from the table.

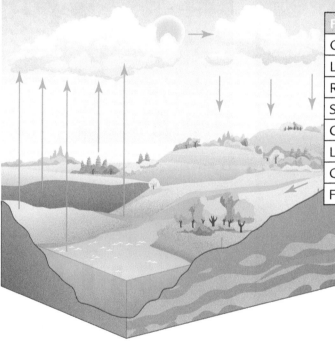

Features	Water movement
Ocean	River flow
Lake	Evaporation
River	Condensation
Sun	Underground flow
Grass	Rainfall
Land	Infiltration
Clouds	Transpiration
Forest	Surface run-off

ISBN: 9780170215701

2 The water cycle is an example of a 'closed system' because no water enters or leaves the system. The system works as a cycle with the water going round in a kind of circle. As the water moves around it changes state. At times it is liquid: in what other states does the water exist in this cycle?

3 The water cycle is not totally closed as a system – there is an input of solar energy from the sun coming into the system. What part would the sun play in the operation of the system?

4 Farming can be seen as another example of a system. Farms are 'open systems' because the farm has inputs coming into it and outputs leaving. These photos show the viticulture system (grape growing).

   a What natural and cultural inputs would go into the vineyard – these will be the things needed for the vines and grapes to grow and then be harvested?

   b Describe the transformations that take place in the vineyard. (What do each of photos A, B and C show?)

   c What are the outputs (photo D) from the vineyard and what happens to these outputs?

ISBN: 9780170215701

Migration: destination Australia

Migration

Migration is a concept as well as a theme that is often the focus of geographic study. The Māori word for this concept is Hekenga.

Migration/hekenga involves the movement of people from one place to another who have the intention of staying permanently or for a long time at the destination. This movement may be to meet the needs of people, or it may be people responding to outside forces like war or natural disasters. The movement of people highlights the way people interact with the natural environment.

New Zealand has a secret region. It is our second largest region after Auckland in terms of population. Where is this secret region? It is Australia, with outlying islands the UK, Canada and the US. More than half a million New Zealand citizens live in these and other countries around the world. And the number is growing.

- Population migration is the name of the process that describes the movement of people from one region or country to another. When people move within a country to live in another part of the country it is called internal migration. When people shift overseas to live it is called external migration.
- In statistics about external migration New Zealand is a stand out. A high percentage of the population living here were born overseas – close to one in every four. These are people who have migrated to New Zealand from overseas and now call this country home. At the same time lots of New Zealanders have emigrated and gone to live overseas. New Zealand and Ireland are the two developed countries with the highest percentage of their citizens (15–20%) now living overseas.
- Australia is, by far, the most popular destination country for Kiwis. 75% of New Zealanders born in New Zealand, but now living overseas live across the Tasman – about 500,000 people. The next largest group live in the United Kingdom (about 60 000). The US and Canada have smaller numbers (around 25 000 and 10 000 respectively).
- Ten times larger: in 1966 there were 50 000 New Zealand born people living in Australia. The figure is now over 500 000.
- In 2010/11 record numbers of New Zealanders have been leaving for Australia – as many as 3000 every month.
- Eight times more New Zealanders live in Australia than there are Australians who live in New Zealand.

Gold Coast

Brisbane

ISBN: 9780170215701

Should John and Leilani move to Australia?

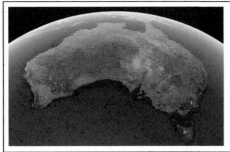  

1. John and Leilani have been married for five years. They have been working in Wellington for a year since they returned from three years OE in Europe.	**2.** Pay rates for nurses and teachers in Australia are 20-30% higher than they are in New Zealand. There seemed to be a shortage of nurses in most parts of Australia.	**3.** Leilani has connections with a local marae in Wellington, is learning te reo and enjoys performing in the cultural group.	**4.** John and Leilani are expecting their first child. They had planned to move to Christchurch and settle there but the earthquakes have made them rethink.
5. Leilani works as a nurse in Wellington. She likes her colleagues but is not happy about the work pressure and staff shortages at the hospital.	**6.** There seem to be few teaching jobs in Brisbane but John was confident he would be able to get work of some kind in Queensland. He has good practical skills and is prepared to do any kind of job.	**7.** Brisbane and other Queensland hospitals are advertising for New Zealand nurses. For taking a contract in inland towns like Mount Isa, and in the tropical north of the state in places like Cairns, cheap housing was being offered.	**8.** John and Lelani live in a small rented apartment in central Wellington. They want to settle down and buy their own home.
9. Air fares between New Zealand and Australia are cheap. Travel across the Tasman is easy for New Zealand passport holders.	**10.** They are both keen on sport. They ski and snowboard in winter, play touch rugby and compete in triathlons.	**11.** John has not been able to get a full time job as a PE teacher. He works as a relief teacher at different Wellington high schools.	**12.** Tropical cyclones from the Coral Sea and recent floods in Queensland worry John and Leilani. They have also heard that climate change could result in water shortages in many parts of Australia in the future.
13. John and Leilani have heard they would not get benefits like health care in Australia until they had lived there for several years. They also read reports about the high cost of houses in the big cities.	**14.** John's parents and grandparents are looking forward to seeing their first grandchild.	**15.** Leilani has a sister and brother living on the Gold Coast. They say the beaches were great all year round and they liked shopping in the many large modern air-conditioned malls.	**16.** Relations living in Australia say they have no regrets about leaving New Zealand and have a higher living standard than they had here. They say there are restaurants catering for all tastes and they have meals out regularly.
17. A few friends have said they found the heat in Queensland unpleasant and they missed the New Zealand change of seasons. They also say New Zealanders are not always welcomed by Queensland locals.			**18.** John is an only child and is very close to his parents and grandparents. They live in Christchurch. Leilani's parents also live in Christchurch.

Fig. 8.4

ISBN: 9780170215701

1 a Write a definition of the term 'population migration'.

b List four pieces of evidence from the case study that proves many New Zealanders have moved/are moving to live overseas?

c Why do you think Australia is the most popular overseas destination for New Zealanders?

2 Migration involves a two-step process: (1) A decision making process; (2) A movement process. Study Figure 8.4 and complete the following tasks.

a Put the information within boxes 1–18 into one of four groups outlined below. You could do this by listing the numbers or by copying and cutting out the 18 boxes and using them as cards to arrange into the four groups.

i Group 1: Background information about John and Leilani

ii Group 2: Reasons to leave New Zealand (push reasons)

iii Group 3: Reasons to move to Queensland (pull reasons)

iv Group 4: Reasons for staying in New Zealand.

b If John and Leilani were your friends and asked you advice about whether or not they should move to Australia, what would you tell them? Give them reasons for your advice.

3 a Draw a map to show Queensland and the locations underlined in the boxes of Figure 8.4.

b What disadvantages can you see for John and Leilani of living and working in a place like Mount Isa as opposed to Brisbane?

c Suppose John and Leilani do decide to move to Australia. Show their 'process of migration' as a flow chart by copying and completing the chart below, adding information in the four frames that would lead on to the end point.

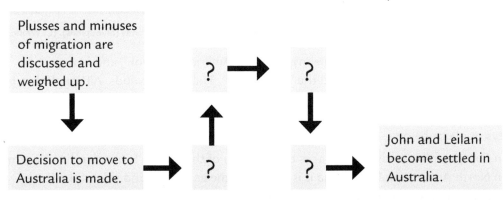

Plusses and minuses of migration are discussed and weighed up.

? → ?

Decision to move to Australia is made. → ? → ? → John and Leilani become settled in Australia.

Mount Isa, an inland mining town.

ISBN: 9780170215701

The Nor'wester

In Māori creation mythology, Tawhirimatea (god of weather) was angry at the separation of his parents (Ranginui, sky father and Papatuanuku, earth mother) by his brothers Tane Mahuta (god of the forests), Tangaroa (god of the sea), Rongomatane (god of cultivated food such as kumara) and Haumia-tikitiki (god of uncultivated food such as the fern root). Tawhirimatea was out to get revenge.

He destroyed forests, caused waves to grow as large as mountains, and the kumara and fern root to burrow deep into Papa, the earth mother. Finally, Tawhirimatea turned on his brother Tumatauenga (god of war and people). But unlike his other brothers, Tumatauenga proved equal to his brother of the winds. Ever since those early times, the weather and people are said to be locked in a continual struggle that neither can win.

Fig. 8.5 Tawhirimatea, god of weather, struggles to control his children, the clouds and storms (blue spirals). (C. Whiting, artist)

In Europe they are called 'föhn' (foehn), in North America 'chinook', and in Argentina 'zonda'. They are sometimes called the 'snow eater'. They are all winds. Although they have different local names they are all the same kind of wind: dry, warm, strong and gusty, that rush down the slopes on the dry and sheltered rain-shadow side of mountains onto the valleys or plains below.

- In Europe, föhn winds affect places to the north of the Alps in Switzerland and Germany. In North America and Argentina it is places on the eastern inland side of the Rocky Mountains and Andes that feel the effects.
- When these winds strike, things warm up quickly. Temperatures can rise dozens of degrees in just a few hours. Föhns can be a pleasant surprise in the middle of winter. People who were just living through below freezing morning conditions can find themselves wearing t-shirts in the afternoon.
- In North America some huge temperature changes have been recorded due the chinook. For instance on 22 January 1943 in Spearfish, South Dakota the temperature went from -20°C to 8.3°C in just two minutes.

ISBN: 9780170215701

The New Zealand Nor'wester

In New Zealand these winds are called 'Nor'westers'. They are usually thought of as Canterbury winds, but in fact occur along the eastern side of both islands. Hawke's Bay, for example, feels the effects of these winds.

They affect both the natural environment and people, impacting on both rural and urban areas. Nor'west winds are famous for:

- bringing very rapid increases in temperature
- causing avalanches, snow melt and floods
- being strong and often gale force in strength
- lifting roofs, snapping branches and uprooting trees
- drying out the land and crops and blowing away topsoil
- causing drought and triggering grassland and forest fires
- causing people to suffer migraines and sleepless nights and become depressed and irritable

People living in Canterbury are familiar with the signs of an approaching Nor'wester. Streaky clouds stretch across the Southern Alps, leaving a gap of clear, light sky against which the mountains and ranges are clearly outlined. Strong winds then begin roaring down the long, wide valleys of the Waimakariri, Rakaia and Rangitata rivers. Heavy rain falls in the upper reaches of these rivers and on the mountains above. Cantabrians know when a Nor'wester is approaching and that they are about to be hit by strong, hot, dry winds that makes conditions unpleasant all the way to the coast.

What causes the Nor'wester?

A one sentence answer to how the natural processes operate would be:
'The Nor'wester (and all föhn winds) occur when a warm moist wind runs into a mountain range, cools slowly as it rises above it, and warms quickly as it descends the other side.'

Air moving across the Tasman towards New Zealand picks up moisture (water vapour) from the sea. This air faces the barrier of the Southern Alps and is forced to rise up and over these mountains on its passage eastwards. As the air rises up it cools down. The higher it rises the more its temperature drops. The already moist air becomes saturated and the moisture condenses out to form clouds. Heavy rain then occurs on the western side of the Alps and across the main divide. The amount of water vapour in the air therefore decreases.

After passing the top of the mountain divide and beginning the descent on the eastern side of the mountain, the air becomes warmer as it loses altitude. Because the air is dry, it gains temperature at a faster rate than it lost when it contained lots of moisture. This results in winds that are drier and warmer on the eastern side of the mountains than they were when they first struck the western side. The eastern side of the mountains are in a rain-shadow area (an area with low rainfall on the inland side of mountains) and get the full effects of the strong, warm and dry winds.

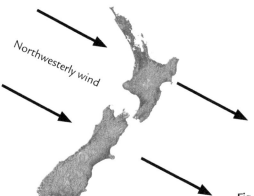

Northwesterly wind

Fig. 8.6 The Nor'wester brings cloud and rain west of the Southern Alps, and strong dry winds on the eastern side

ISBN: 9780170215701

Some Nor'wester statistics

The key to the Nor'wester process (and all föhn winds) is the release of latent heat.
Latent heat is like a hidden or dormant heat that only escapes and becomes obvious under certain conditions. It is released when moist air rises over hills and mountains. This heat has a big effect on the dry descending air on the other side of the mountains.

- As the air rises it cools, losing heat at a rate of 1 °C for every 100 m increase in altitude.
- If there is enough moisture in the rising air, the water vapour in the air condenses into water droplets to form clouds. This happens when the air becomes fully saturated with moisture.
- In the process of changing from water vapour into water droplets, latent heat is released.
- This extra heat reduces the rate of cooling of the air. The rate of cooling drops to between 0.5 and 0.65 °C per 100 m.
- The rising air has therefore cooled at two different rates: a fast rate and a slower rate.
- Once over the mountains the air is drier because much of the water has been lost from it as rain.
- This air warms up as it descends from the mountains.
- The air warms at a rate of 1 °C for every 100 m loss of height (there is less water in it to heat). The air warms at this same fast rate all the way down until it reaches the plains.
- It ends up at sea level on the east coast of the South Island much warmer than it was when it began its journey across the island on the West Coast.

- Air can arrive on the West Coast at 15 °C, it rises and cools to 3 °C at the top of the Southern Alps, it then descends and warms up faster than it cooled down and reaches the Canterbury Plains, bringing temperatures of up to 27 °C to the coast.
- As well as causing temperature change, the Nor'wester effect is also related to rainfall variation between the west and east side of the Southern Alps through orographic rainfall and rain shadow. This helps explain why the west is much wetter than the east.
- In a single 24-hour period in November 2006, over 300 mm of rain fell in the Cropp Valley on the western side of the Southern Alps. In that same 24 hours, that same wind brought a dry, mild spring day to Christchurch and no rain at all.

West of the alps	Time and temperature	East of the alps	Time and temperature
Hokitika	3 pm: 15 °C 4 pm: 14 °C 5 pm: 14 °C 6 pm: 14 °C	Ashburton	3 pm: 20 °C 4 pm: 24 °C 5 pm: 25 °C 6 pm: 23 °C

Table 8.1 Weather reports from Hokitika and Ashburton for 16 October 2010, showing effect of the Nor'wester on temperature

ISBN: 9780170215701

ACTIVITY

1 a Make a copy of the Nor'wester formation process diagram in Figure 8.7 (below), and add the following labels:

West   Southern Alps   Canterbury Plains   The Nor'wester
Tasman Sea   Pacific Ocean   East

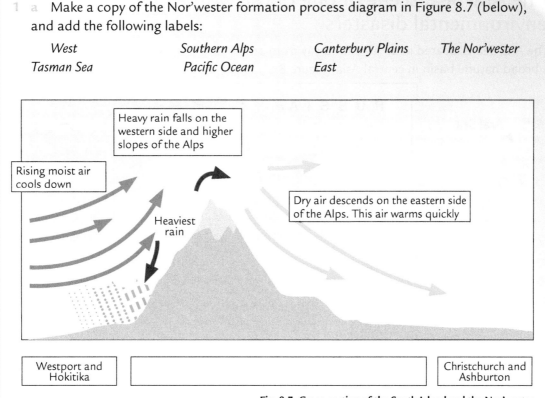

Heavy rain falls on the western side and higher slopes of the Alps

Rising moist air cools down

Heaviest rain

Dry air descends on the eastern side of the Alps. This air warms quickly

Westport and Hokitika

Christchurch and Ashburton

Fig. 8.7 Cross-section of the South Island and the Nor'wester

b Complete Table 8.2 (below) and describe the positive, negative and interesting features of the Nor'wester or föhn winds.

Features of Nor'wester/föhn winds		
Positive	**Negative**	**Interesting**
1.	1.	1.

Table 8.2 Positive, negative and interesting features of the Nor'wester and föhn winds

2 a If the outcome of a natural process is the hot, dry Nor'wester across the Canterbury Plains, describe and explain the natural events that lead to this outcome.

   b Refer to Table 8.1 and construct a pair of bar graphs or thermometers to compare and contrast the temperature in Hokitika with that in Ashburton at 5 pm on 16 October 2010.

3 Either write a paragraph to explain or make a drawing to illustrate what is meant by the idea in the Māori creation myth that 'the weather and people have been locked in a continual struggle that neither can win'.

ISBN: 9780170215701

The Aral Sea: scene of one of the world's worst environmental disasters

The Aral Sea is located in an area far away from any ocean. The sea is really a lake formed in a broad natural basin in central Asia (Figure 8.8). It has no outflow to the ocean.

Fig. 8.8 The Aral Sea in central Asia

Natural processes have led to the formation of the Aral. It is part of a water cycle: snowmelt and rain run-off from the Tien Shan and Pamir mountains feed into the rivers Syr Darya and Amu Darya (Figure 8.9). These rivers flow 2000 km northwest from the mountains into and through a desert and semi-desert area. The water from the rivers flows into a depression resulting in the formation of a large lake (the Aral Sea).

Evaporation removes water from the sea thereby balancing water inputs (from the rivers) with water outputs (from evaporation). This evaporated water feeds back into the atmosphere adding to the moisture that falls on the far away mountains as rain and snow.

Fig. 8.9 Tien Shan mountains: rain and spring snowmelt feed the rivers that flow into the Aral Sea

ISBN: 9780170215701

History

In the 1950s and 60s government planners in Moscow had what seemed like a good idea. Central Asia was an area with a large amount of flat land, low population density and low agricultural production. The government wanted increased national production of essential food and industrial crops like rice and cotton, and used water from the Syr Darya and Amu Darya rivers to irrigate huge areas of land between the mountains to the south-east and the Aral Sea to grow crops.

Some irrigation had always been practised on fertile land close to the rivers but nothing on the vast scale of what to take place over the next 30 years (Figure 8.10). Canals were built to divert river water. Eight million hectares of land were irrigated. The government set production goals for farm output. From a small, diversified crop output to meet local needs there was now a virtual monoculture of mass production focusing on cotton and rice (Figure 8.11).

The Aral area is in a latitude band similar to that of southern New Zealand but it has a very different climate. The climate is 'continental', which means features of inland locations that are far away from oceans. These places usually have extreme weather conditions with hot summers, very cold winters and low rainfall. The Aral area has summers with temperatures regularly in excess of 30 °C and winters with temperatures dropping to -10 °C. Annual rainfall in the area is less than 250 mm.

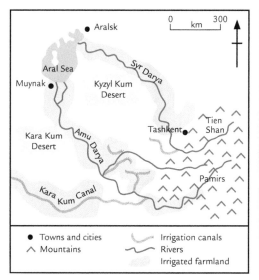 

Fig. 8.10 Central Asia: irrigation schemes (left) and an irrigation canal from the Syr Darya river in Kazakhstan (right)

Fig. 8.11 Irrigated onion (left) and cotton crops (right) in Uzbekistan

ISBN: 9780170215701

Fast forward to the 1990s

'Aral' means 'Sea of Islands', a name that seems ever more appropriate since more islands continued to appear as the level of the sea fell. Cotton production ('white gold') was certainly high but at great cost to the environment. Irrigation had taken away the lifeblood of the sea. Rivers feeding the sea were reduced to a trickle or failed to reach the sea at all. Fishing towns like Aralsk that had once been on the shores of the sea now stood high and dry with the sea shore hundreds of kilometres and several hours drive away.

The Aral, once the world's fourth largest lake, was now only one tenth of its former size and had split into four lakes (Figure 8.12). Continued evaporation and reduced amount of water meant a build-up of salt and other chemicals. The Aral became a virtual dead sea with few fish species able to survive. Not only were fishing boats left stranded on a drying sea bed, but the fish that the boats once caught were no longer present. The Aral based fishing industry of towns like Muynak – catching, processing, canning and exporting fish – could not survive. More than 60 000 people lost their jobs.

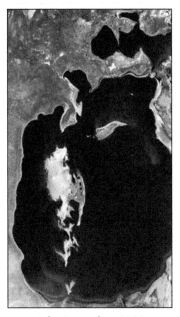
July–September, 1989

August 12, 2003

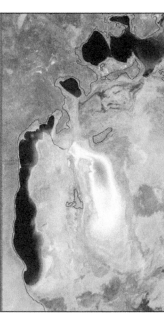
August 16, 2009

Fig. 8.12 Satellite imagery showing changes in the Aral Sea from 1989 to 2009

The environment of the Aral basin also deteriorated (Figures 8.13 and 8.14). With less water in the sea there is less evaporation and hence less rainfall. Salt dust storms increased. Locally these storms were known as the 'dry tears of the Aral'. Heavy use of fertiliser and pesticide increased to maintain production, coupled with poor farming practices on the irrigated farmlands, led to chemical rich run-off back into the rivers. The salt, chemicals and pesticides created a toxic mix in the rivers, lakes and in the Aral itself.

Thirty-five million people live in this central Asian region, 1.3 million around the Aral Sea alone. Significant health deterioration has been recorded: infant death rates, throat cancer, respiratory and eye diseases have all increased. Health decline has been linked with the deterioration of the environment. Traces of pesticide have been found in mothers' breast milk.

ISBN: 9780170215701

Fig. 8.13 The former fishing port of Aralsk. Where the calf and cow graze, there used to be water. Ships once unloaded by dockside cranes, but have now become stranded on the lake bed unable to escape, abandoned and left to rust

The irrigation schemes were the product of the USSR, and since they were introduced the USSR has been broken up into a number of new countries. The Aral Sea and basin are now mostly contained within the borders of Kazakhstan, Uzbekistan and Turkmenistan. These countries have inherited the problems of the Aral. In Uzbekistan people posted signs 'Forgive us Aral. Please come back' and 'The Aral will live again'. The people live in hope.

Fig. 8.14 Shrinking sea, expanding desert, stranded, abandoned and rusting ships.

ISBN: 9780170215701

The situation in 2012

There is hope that at least part of the Aral Sea can be saved. Kazakhstan, using wealth from its vast oil fields, has put money into saving the northern part of the Aral. With World Bank help, irrigation methods have been improved to take less water from the Syr Darya river and to make better farm use of the water that is taken. Water flow into the northern part of the Aral from the Syr Darya river has increased. The 13-km-long Kok-Aral dam has been built to isolate the northern part of the Aral from the southern part so that water inflow from the Syr Darya is retained in the north.

Native plants, birds and, most important of all, fish (pike, perch and carp) have returned to the growing sea. Aralsk is now seeing water reappear again and there is hope for a return of a fishing fleet. Middle-aged men and women who left the region when they were young are starting to return for the fishing, and they are building houses. Billboards announcing, 'Good News — the Sea is Coming Back' stand beside new hotels, and some small-scale fish processing has begun again.

The southern sea, which lies in poorer Uzbekistan, has mostly been abandoned to its fate and is still shrinking. Uzbekistan is one of the world's top five cotton producing countries and is the largest exporter of the material. The country shows little interest in reducing cotton irrigation to help restore water flow into the Amu Darya river and then into the Aral. Instead, Uzbekistan is moving toward oil exploration in the drying South Aral seabed. The European Space Agency expects the southern Aral to dry out completely by 2020.

ACTIVITY

1   a   Match the names and features in Table 8.3.

Name	Feature
1. Aralsk	A. River that flows from the Tien Shan mountains into the Aral Sea.
2. Kok-Aral	B. Oil rich central Asian country that is helping the Aral Sea recover.
3. Syr Darya	C. Long irrigation canal taking water from the Amu Darya river into Turkmenistan.
4. Uzbekistan	D. River that has had so much water taken from it that it no longer flows into the Aral Sea.
5. Pamirs	E. Central Asia desert that covers large parts of Kazakhstan and Uzbekistan.
6. Kara Kum	F. High mountain range, the source of the Amu Darya.
7. Tashkent	G. Dam that is helping restore the northern part of the Aral Sea.
8. Kyzyl Kum	H. Capital city of Uzbekistan.
9. Kazakhstan	I. Former fishing port that was once on the shore of the Aral Sea in Kazakhstan.
10. Amu Darya	J. Country using a huge amount water to irrigate crops of cotton.

Table 8.3 Natural features and place names of the Aral region

b   Either draw and annotate a map or write a paragraph to highlight important features of the Aral Sea.

c   Why were the ships in Figures 8.13 and 8.14 not taken into deeper ocean water when the Aral began to shrink?

ISBN: 9780170215701

d The diagram below shows how the natural water cycle in the Aral area used to work. Copy and complete the diagram by adding information into boxes A and C.

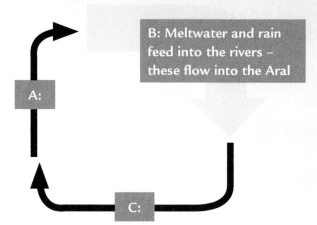

B: Meltwater and rain feed into the rivers – these flow into the Aral

A:

C:

2 a Find the meaning for each of the words below. Include specific information relevant to the Aral in your definitions. They are listed in the order they appear in the text:

1 run-off   2 evaporation   3 irrigation   4 latitude

5 continental climate   6 lifeblood   7 stranded   8 toxic

9 isolate   10 billboard

b Table 8.4 lists the world's five largest lakes as well as several other lakes around the world. Draw a graph of an appropriate type to illustrate these statistics.

Lake name	Surface area (sq/km)
Caspian Sea (Central Asia)	372 000
Lake Superior (North America)	82 000
Lake Victoria (Africa)	70 000
Lake Huron (North America)	60 000
Lake Michigan (North America)	58 000
Old Aral Sea (1950)	68 000
Aral Sea (2008)	9000
Lake Taupo	606

Table 8.4 Lake surface area sizes

c Provide captions for the three satellite images in Figure 8.12.

3 a Choose two contrasting landscape photos from Figures 8.9, 8.10, 8.11, 8.13 and 8.14. Draw annotated sketches of the two photos you have chosen and title each.

b Provide a full analysis (description and explanation) of how irrigation impacted on the Aral basin area. Refer to the sequence of events that took place and both the natural and cultural environment in your answer.

ISBN: 9780170215701

ISBN 9780170215701

Spatial concepts

9

Location
Distance
Accessibility

Ancient and modern meet on the
Great Silk Road in Mongolia.

ISBN: 9780170215701

GEOGRAPHY DICTIONARY

Spatial is a word related to how features are arranged on the earth's surface. The word 'spatial' means 'related to space'. In geography, 'space' refers to the surface of the earth. The places and areas that make up the surface of the earth take up the earth's 'space'. Geography describes and explains this 'earth space'. An easy way of understanding the word 'spatial' is to think of it being connected with 'place' (in contrast to the word 'temporal' which is time-related)

The word 'spatial' is not commonly used. Even in geography books its use is rare, however, studies in geography almost always involve spatial concepts of one sort or another. All of the following terms or concepts have a spatial focus or connection:

- Location
- Place
- Area
- Scale
- Distribution
- Region
- Land use
- Distance
- Accessibility
- Direction

When you read a list like this it is clear that spatial studies do not need to involve the actual word itself. Geography has been called a 'spatial subject' because it involves describing and explaining places, and thereby makes frequent use of spatial concepts. If we broaden the meaning of the word 'space' to mean areas and places, then spatial studies and spatial understanding can be seen as an important part of any geographic study.

Figure 9.1 shows an image of Earth that has been made using satellite data and transformed into a two-dimensional map format. It is an image taken from outer space, and highlights how water covers more of the earth's space (surface) than land. The different land surface colours shows things like deserts (pale brown), snow and ice cover (white) or vegetated areas (green).

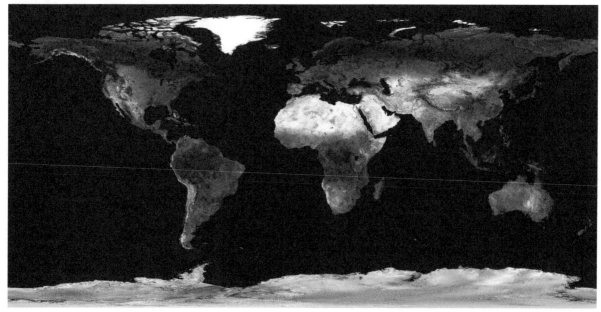

Fig. 9.1 The Earth: an image made using satellite information

ISBN: 9780170215701

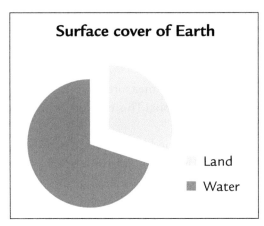

Surface cover of Earth

Land

Water

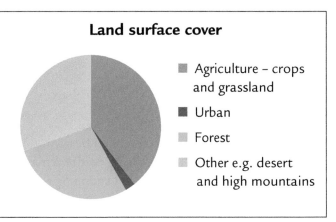

Land surface cover

- Agriculture – crops and grassland
- Urban
- Forest
- Other e.g. desert and high mountains

Fig. 9.2 The surface of the earth

ACTIVITY

1 a Unscramble these spatial related words. Two letters have been reversed in each word in the grid.

S	C	A	P	E					
P	L	E	C	A					
A	C	T	E	S	S	I	B	I	L	I	C	Y
T	E	A	R	R	I	N			
I	A	O	L	S	T	E	D		
A	R	A	E						
L	T	C	A	O	I	O	N		

  b Use each of the unscrambled words above to write a short geography-related statement to show the meaning of that term. Example: *People are interested in where land judged no longer safe to build on (red zone land) in Christchurch after the 2010 and 2011 earthquakes is located. This is a spatial study.*

2 Write a paragraph describing the surface cover of the earth. Refer to Figures 9.1 and 9.2 in your answer.

3 a Convert the information from Figure 9.2 into bar graphs.

  b What surface land cover would you expect to find in:

    i   Urban areas?   ii   High mountains?

ISBN: 9780170215701

CASE STUDY 22

Location, distance and accessibility

Location: is about the position of places. It can be described and measured in different ways. Latitude, longitude and grid positions can give precise locations. This type of information is used in GPS units and by Google Earth maps.

Location can also be described by saying how a place is positioned in relation to other features or phenomenon. Being next to a main road, at a road junction, near to a beach, across the road from a park or shopping centre, or at the foot of a mountain range all describe the locations of places.

Distance: can be measured in units of length like metres and kilometres. This type of distance measurement is valuable but has become increasingly less important than 'travel time' as a measure of distance. Travel time is a measure of how long it takes to travel between two places, so how far apart places are in travel time is another way of describing distance. 'Time distance' or speed of travel becomes another measure of distance alongside the actual 'kilometre distance'. This idea can be taken further to include other ways of measuring distance like calculating the 'cost of travel', for example the cost of petrol consumption by a car, or in air fare costs.

Fig. 9.3 Location and distance signage

Destination	Distance from Auckland (km)	Direct flight time (hours and minutes)	Cost per person one way (NZ$)
Dubai	14 175	18 h 17 m	1389
Sydney	2140	2 h 46 m	218
Buenos Aires	10 450	11 h 50 m	2112
Rarotonga	3000	4 h 05 m	472
Los Angeles	10 500	13 h 30 m	1310
San Francisco	10 490	13 h 32 m	1399
London	18 320	23 h 40 m	1446
Hong Kong	9092	11 h 44 m	774
Melbourne	2642	3 h 25 m	265
Brisbane	2260	2 h 55 m	227

Table 9.1 Travel distance, times and cost from Auckland

Accessibility: on the other hand, is a measure of how easy it is to travel or move between places. Distance and location both affect accessibility: some places are easier than others to get to.

In the modern world, movement between places could perhaps better be described as 'connections' between places. Satellites and the Internet provide 'instant connections and communications' between places, but just as kilometre, time and cost factors make some places more accessible than others, places without cell phone coverage or other communications technology remain outside of the Internet- and satellite-connected world.

Departures   17:59

DESTINATION	BOARDING TIME	GATE	STATUS
DUBAI	17:35	15	FINAL CALL
SYDNEY	17:50	04	BOARDING
SYDNEY	18:05	16	BOARDING IN 5 MINS
DUBAI	18:10	02	BOARDING IN 10 MINS
BUENOS AIRES	18:25	07	BOARDING IN 25 MINS
RAROTONGA	18:35	01	BOARDING IN 35 MINS
LOS ANGELES	18:35	06	BOARDING IN 35 MINS
SAN FRANCISCO	18:50		BOARDING IN 50 MINS
LOS ANGELES	20:10		PLEASE WAIT
RAROTONGA	21:05		
LONDON	22:05		
HONG KONG	23:15		
SYDNEY	05:05		
SYDNEY	05:20		
MELBOURNE	05:30		
BRISBANE	05:50		
SYDNEY	05:50		

Please do not leave baggage unattended.

Fig. 9.4 Accessibility signage

Geographer Nick Middleton was researching his new book and making a TV series that involved travel along the Great Silk Road, an ancient trading route running for thousands of kilometres between Xian, central China and the Mediterranean Sea in Europe.

He described an experience in Mongolia, part way along the road: 'I was struck by two things. The nomads way of life has in many ways changed little since days of Genghis Khan (13th century AD) – they still herd animals and live in round felt tents (yurts) – but at the same time modern influences have seeped into their daily routine. Virtually every yurt I came across was adorned with solar panels that powered an electric light and, in some cases, a television set permanently tuned to wrestling tournaments beamed from Ulan Bator (the capital city of Mongolia)'.

ISBN: 9780170215701

Fig. 9.5 The meeting of tradition and technology on the Great Silk Road

1 a Refer to the departures board from Auckland International Airport in Figure 9.4. Use an atlas and a world map to mark all the destination locations shown. Draw linkage lines from Auckland to each of these destination locations. Draw multiple lines if there is more than one flight connection.

   b Create a table of the flight destinations shown on the departures board. In the table record the latitude and longitude of each place (use an atlas or Google Earth to find this information). Here is a start:

Destination city	Latitude	Longitude
Dubai	25°16' North	55°19' East

   c There are 17 flights due to depart.
      i   How many are going to Australian destinations?
      ii Why would there be more flights to Australian destinations than to any other destination?

   d Referring to Table 9.1, which of the distance measures shown would the following travellers be most interested in or concerned about? Give reasons for your answers.
      i   A chief executive of a company travelling on business.
      ii   A sports team travelling to a tournament.
      iii A teacher travelling to a conference.
      iv   A family of five going on vacation.

   e Imagine you could travel to any one of the destinations in Figure 9.4. Where would you choose to go and why?

   f What other flight information or details would you want before making a booking to your chosen destination?

ACTIVITY

ISBN: 9780170215701

2 a Make a copy of the map below. Use your geography understanding and a little imagination to create a key for all the map symbols and colours. (Hint: there is no 'correct answer' – it is your choice.) Give the map a key, scale and title.

Fig. 9.6 An imaginary area

b Describe what the region you have created is like.

c Put a letter 'A' on your map to show which place/location is the most accessible. Justify your choice, explaining what makes this place more accessible than others on the map.

3 a Refer to Nick Middleton's account and Figure 9.5 to answer the following questions.

  i Use an atlas and describe the location of Mongolia. Include a range of ways that can be used to the describe location.

  ii What has helped break the barrier of distance and remoteness for nomads living in this area?

b Detective work: looking at the signpost in Figure 9.3, can you work out where in the world this signpost is located? Write down the steps you took to arrive at your answer.

ISBN: 9780170215701

ISBN: 9780170215701

CASE STUDY 23

Location, location, location

Spatial information is data that is linked to a geographic location. It has thousands of uses, making it possible to do things like use maps on mobile phones or send emergency services to the right addresses. Spatial information is part of our everyday lives, so much so that we don't even think about it. Without this data, however, modern society would grind to a halt.

Human activity depends on spatial information, on knowing where things are and understanding how they relate to one another. Having access to spatial information enables key service providers to answer 'Where?' questions in a range of critical situations:

- Where is a distress call coming from?
- Where is a fire or severe weather front moving to?
- Where is the source of a flu epidemic and how far has it spread?
- Where are troops in the combat zone? Where should we target a missile attack on the enemy?

Then there are the uses for business, government and community:

- Aerial photographs can be used to construct a 3D neighbourhood model to illustrate advantages and disadvantages about a proposed shopping mall development.
- Businesses can plot the home locations of customers on a map to help target marketing and delivery of products.
- Software companies use data, maps and photos of real cities to help construct virtual cities to be incorporated in computer games.
- Security agencies use building footprints and heights to plan protective services for state visits (to identify possible hiding places for attackers and for locations where security agents could be placed), as well as for security exercises such as plotting plume (smoke) overlays to estimate spread and impact of debris following a potential terrorist attack.
- Creating graphics to simulate real time images of sports events such as yacht races or golf tournaments.

'Beautifully restored old two story villa for sale. Located in a quiet tree-lined street in a neighbourhood of similar high quality homes. There is a park opposite, and the village cafes, library and shops are just a five minute stroll away. The area is in the local grammar school zone. It is just a five minute drive to the motorway. The railway station with trains every half hour into the city is within easy walking distance. Be quick as properties like this don't last!'

**Open home tomorrow
1–2 pm.**

Fig. 9.7 House for sale: location can be the selling point!

ACTIVITY

1 a Draw a star diagram to show how we use spatial data and spatial information.

b Explain why this kind of data is so important in today's world.

2 List the advantageous and positive location features of the house for sale in Figure 9.7.

3 Study the photograph of the property below, then create a real estate advertisement to help sell it. Highlight attractive locational features of the property. Include a sketch as a part of your advert.

Fig. 9.8 For Sale

ISBN: 9780170215701

CASE STUDY 24

Antipodes

Place a needle on the surface of the earth and push it through the surface. Make sure the needle passes directly through the centre of the interior, and continue pushing until the needle punches another hole out through the surface on the other side. The entry and exit point of the needle will be on exact opposite sides of the earth. This may be impossible to do in real life but the theory holds true: every place on Earth has a 'twin' that is located on the exact opposite side of the planet. These two points are said to be 'antipodal', meaning that they are the antipodes of each other. Think of the 'antipodes' as 'directly opposite locations'.

South of New Zealand lie the Antipodes Islands. These sub-Antarctic islands are part of New Zealand. They are small, cold and bleak, located about 900 km south of Invercargill. The islands got their name after early British explorers thought the land to be the point nearest to the antipodes of London. (In reality, the exact antipodes of London is in the ocean nearby.)

Looking out from New Zealand, the antipodes of the country is located in the northern hemisphere, across the very north of Africa through Spain and out into the Atlantic Ocean between the coast of France and Spain.

Places that are antipodal in location share other opposite features: midday at one place is midnight at the other; the longest day at one place corresponds to the shortest day at the other; and midwinter at one place coincides with midsummer at the other.

The western hemisphere (Europe, Africa, Asia and Oceania) in Figure 9.9 has a 'normal appearance'. The yellow continent overlay shows antipodes locations (the orange is where two land areas are antipodal). New Zealand shows as having Spain as its antipodes. Across northern Africa (the Sahara desert) the yellow specks reveal Pacific Island countries in their antipodean locations.

Fig. 9.9 Antipodal locations around the world

Some antipodean pairs (or very near matches):

- The most famous of all are the North and South Poles
- Christchurch = La Coruña (Spain)
- Hamilton = Córdoba (Spain)
- Tauranga = Jaén (Spain)
- Whangarei = Tangier (Morocco)
- Auckland = Seville and Málaga (Andalusia, Spain)
- Wellington = Madrid (Spain)
- Campbell Islands (New Zealand) = Limerick (Ireland)
- Te Arai Beach (85 km north of Auckland) = Gibraltar
- Suva (Fiji) = Timbuktu (Mali)

ISBN: 9780170215701

ACTIVITY ▸

1 Use an atlas to find the Antiopdes Islands.

  a Describe their location.

  b Explain how the islands got their name.

2 Referring to Figure 9.9 and an atlas, what places are antipodal to:

  a Greenland?

  b The US?

  c Argentina?

3 Sketch a globe and add two labels on antipodal sides as shown below. For label A, write the name of where you live in New Zealand. For label B write the name of the antipodal location. This may require some research. (Hint: the antipodes pairs listed previously might help.)

ISBN: 9780170215701

Sustainability

10

Amazonia: using rainforests in a
sustainable way is a major challenge.

ISBN: 9780170215701

GEOGRAPHY DICTIONARY

Sustainability involves adopting ways of thinking or behaviour that allow individuals, groups and societies to meet their needs and aspirations without preventing future generations from meeting theirs. Sustainable interaction with the environment may be achieved by preventing, limiting, minimising or correcting environmental damage to water, air and soil as well as considering ecosystems and problems related to waste, noise and visual pollution.

Fig. 10.1

Part 1· Sustainability and planet Earth

Definitions

- The word sustainability is derived from the Latin word *sustinere* (*tenere* means 'to hold' and *sus* means 'up'). Dictionaries state that the word 'sustain' means to 'maintain or support'.

- One commonly used definition of sustainability is, *'Improving the quality of human life while living within the carrying capacity of supporting ecosystems and for this to be able to continue into the future'*.

- Sustainability is about how we live now and whether this lifestyle can be maintained and improved on in the future. The term is connected with the impacts of the human lifestyle on natural resources and the natural environment, but it is also linked with the social and economic impacts of how we live.

ISBN: 9780170215701

- Sustainability means striking a balance between using the resources and processes of nature and the ability of nature to replace or treat the effects of that use. Sustainability relates to the earth's capabilities. The healthier Earth's systems are, the more resources there are to use, but with damage to the Earth's systems, resources for everyone are reduced.

Most people agree that sustainability is important, however, it is not an easy concept to define in a way that everyone agrees with. There are many different definitions given of the word. Despite this most definitions of sustainability include reference to these five factors:

A   The way people live today (our lifestyle).

B   How people use the environment and the earth's resources to support and maintain their lifestyle.

C   The social and economic impacts of the way people live.

D   Whether or not people can live as they do in the present day on an ongoing basis.

E   How people of the future might live and be able to live.

These five things are interconnected. The 'people' in these statements refer to you and I, as individuals and in our communities. We are all involved in the issue of sustainability. The way we live today and the way we will live in the future are all affected by and impact on sustainability.

Other ways of looking at sustainability: sustainability pairings
Sustainability can be used as a word on its own but is often paired with other words. Here are some examples of these pairings:

- Sustainable farming
- Sustainable development
- Sustainable tourism
- Sustainable environment

- Sustainable water use
- Sustainable lifestyle
- Sustainable cities

ISBN: 9780170215701

1 a Using a dictionary to help you, rearrange Table 10.1 below so that the words and definitions are correctly matched.

Sustainability terms	
Societies	Carrying on with something and continuing to do it.
Aspirations	Community of living things and their relationship with their environment.
Interaction	Somebody or something that is of use and value.
Ecosystems	Communities and groups of people that have some common connections.
Fragile	Having the desire to do and achieve things.
Resources	Living and doing things in ways that can be maintained.
Ongoing	Two or more things having an effect on each other and working together.
Sustainability	Easily damaged or easily broken.

Table 10.1

b List the sustainability pairings on page 141 in alphabetical order.

c Hidden within the following text is a message about sustainability, however, there are no word breaks. Work out where the spaces between each word would be and rewrite the message as a single sentence. Starting where the arrow points, read from right to left along the top line, continuing left to right along the middle line and then finish the bottom line by reading from right to left.

I W Y T I L I B A N I A T S U S G N I V E I H C A ←

L L L E T T H E E A R T H C O N T I N U E S U P P

T I W O N K E W S A E F I L N A M U H G N I T R O

2 Which of the sustainability pairings listed previously do you think is the most important? Give a reason for your choice.

3 a Make a copy of Figure 10.1. Label the names of the continents, seas and oceans that are shown.

b Explain the message in this picture (Figure 10.1). Include the word 'sustainability' in your answer (check you know the meaning of the word 'fragile').

Part 2 · Sustainability from different perspectives

Perspective 1:
The Earth is in our hands

Fig. 10.2

ISBN: 9780170215701

Perspective 2

Cree Indian prophecy: *'Only after the last tree has been cut down; only after the last river has been poisoned; only after the last fish has been caught; only then will you find that money cannot be eaten.'*

Perspective 3

Herman Daly, environmentalist and economist:
'What use is a saw mill without a forest?'

Perspective 4

Sir David Attenborough, naturalist and broadcaster:
'Instead of controlling the environment for the benefit of the population, perhaps we should control the population to ensure the survival of our environment.'

Perspective 5

Christopher Flavin, Worldwatch Institute president: *'Building a world where we meet our own needs without denying future generations a healthy society is not impossible, as some would assert. The question is where societies choose to put their creative efforts.'*

Perspective 6

World leaders at the summit on Sustainable Development stated this aim: *'To free all of humanity, and above all our children and grandchildren, from the threat of living on a planet irredeemably spoilt by human activities, and whose resources would no longer be sufficient for their needs.'*

Perspective 7

Paul Rose, explorer, television presenter and vice president of the Royal Geographical Society: *'I am very optimistic about our future sustainable development. Many of us now have the will and technology to protect our ecosystems and improve the way we all live. It's an awesome responsibility and we can do it. If we act smartly, I believe that in the next 30 years we will have drastically slowed down the human effects on our climate and made sensible, sustainable adaptations to the changes that are happening now, have reduced our rate of population growth to sustainable level and have protected our ecosystems and biodiversity.'*

Perspective 8

The Gifts of Life and Pollution
Nature's gifts are sunny beaches,
Nature's gifts are rocky mountains,
Nature's gifts are vast oceans,
Nature's gifts are the colours in the rainbow,
Nature's gifts are rain forests,
Nature's gift is life.

Our gifts are oceans filled with toxins,
Our gifts are global warming,
Our gifts are deserted forests,
Our gifts are killed animals,
Our gifts are serious diseases,
Our gift is pollution.

Perspective 9

ISBN: 9780170215701

ACTIVITY

1 Refer to the images and different perspectives 1–9 on pages 142-143 to answer the following questions.

   a Figure 10.2 shows the world in the hands of people (our hands). List three things suggested in Perspectives 2, 3 and 4 that we could do to make a better and more sustainable world.

2 What type of world do the leaders in Perspective 6 want their children and grandchildren to live in?

3 Paul Rose (Perspective 7) has an optimistic view of the future. What makes him optimistic?

4   a Who or what gives the earth life and who or what causes pollution according to the student's poem (Perspective 8)?

   b Draw a half or full page circle to represent the world. In the left side provide drawings that illustrate the 'gifts of nature' part (first stanza) of the poem, and in the right side illustrate 'our gifts' (second stanza).

Part 3 · Sustainability involves society and the economy as much as it does the environment

The diagrams below (Figures 10.3 and 10.4) suggest that sustainability of the planet involves three sets of factors: environmental, economic and social. Although people may argue that any one of these factors is all-important and underpins the others, a stronger case can be made for saying that all three work together (complement each other) and are equally important.

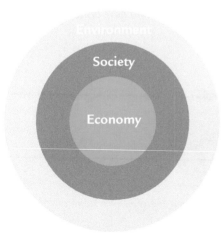

Fig. 10.3

Fig. 10.4

ISBN: 9780170215701

ACTIVITY

1 a Sustainability of communities and of the earth is connected by three things. Refer to Figure 10.3 and state what these things are.

   b How does the diagram show that these things are connected?

   c Write out a copy of Table 10.2 below and write the word 'social', 'environmental' or 'economic' in box A, B or C to correctly categorise each statement.

Types of sustainability: social, environmental, economic
A Polluted water from the town and farms that ran into the lake made the lake unable to support any life. All the fish died. The lake became black and smelly. It was a dead lake. People avoided the lake and wondered how things could have gone so wrong and if there was any way back.
B When the mines closed the future of the town looked bleak. People had to move away to find jobs. The community came up with the idea of developing and promoting the area as a centre for adventure tourism. Bike trails were built and challenge courses were set up in the old mine sites. Tourists started to come. The area seemed to have turned a corner towards developing a stable and prosperous community based on tourism.
C The signs were there. Small protests took place every day about lack of freedom, and people being denied their rights to fair and equal treatment. The police and army were in control but it was a situation that could not last. Once the mass protests began the situation became unsustainable and the government was overthrown.

Table 10.2

2 a Copy Figure 10.4 and rewrite the three labels in your own words.

   b Many people believe that global sustainability must start with the environment. How does Figure 10.3 emphasise that the environment is all important for a sustainable planet?

   c Look carefully at the sustainability logo below (Figure 10.5). Mention a possible reason for the use of colour and other design features.

Fig. 10.5 This black and white logo has been used to represent the idea of sustainability

   d What do you think the designers are trying to say with this logo?

3 Many businesses promote the idea that they operate in sustainable ways. Design a logo that a company could use on their packaging, letterhead and web site that emphasise they are a 'sustainable enterprise'.

ISBN: 9780170215701

CASE STUDY 25

Tongariro National Park: environment, people and sustainability

History

In 1887, Te Heuheu Tukino IV (Horonuku), supreme chief of Ngati Tuwharetoa, gifted the sacred peaks Tongariro, Ngauruhoe and Ruapehu to the nation. This land marked the beginnings of the Tongariro National Park. It became New Zealand's first national park.

Ngati Tuwharetoa and Ngati Rangi have lived beneath the mountains for centuries. The mountains are a part of their history, their whakapapa (genealogy) and legends. The mountains are looked on with reverence and respect, and as spiritual places that have the ability to both create and to destroy life on a huge scale. For tangata whenua the mountains are ancestors: where the people have come from and where they will return to. The mountains are tapu (a sacred place).

There is a story about the ancestor Ngatoroirangi (the navigator and tohunga of the waka *Arawa*). He was cold and close to death after exploring this mountainous region, and so called out to his sisters from his Pacific homeland, Hawaiiki, to send him fire. The fire came, but its journey left a trail of volcanic vents from Tongatapu, through Whakaari (White Island), Rotorua and Tokaanu, before reaching Ngatoroirangi on the slopes of Tongariro.

The place: location and size

- Part of the Volcanic Plateau region of the central North Island.
- Located between latitude 38° 58' to 39° 25' South, and longitude 175° 22 to 175° 48' East.
- Located southwest of Lake Taupo.
- Park area size is 80 000 hectares (Lake Taupo surface area is 62 000 hectares).
- Towns of Turangi, Waiouru, National Park and Ohakune are on the edge of the park.
- Main roads run around the park boundaries.

KEY
A Lake Taupo
B Turangi
C Lake Rotoaira
D Tongariro Alpine Crossing area
E Mt Tongariro
F Mt Ngauruhoe
G National Park township
H Tama Lakes
I Whakapapa skifield
J Mt Ruapehu
K Turoa Skifield
L Ohakune
M Lake Moawhango
· Lakes are indigo/black
· Snow covered areas are white
· Rock and other bare ground surfaces are purple/brown
· Forest, scrub and pasture are different tones of green

Fig. 10.6 Tongariro National Park aerial view

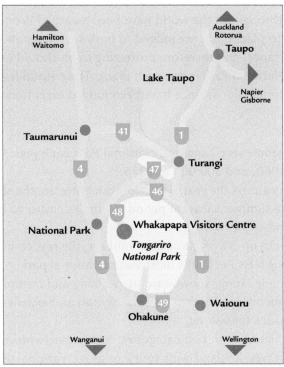

Location map for Tongariro National Park

Fig. 10.7

Geography and geology

- Three active andesitic volcanoes – Tongariro (1968 m), Ngauruhoe (2290 m) and Ruapehu (2797 m) – are at the centre of the park.
- Ruapehu is the highest North Island mountain and includes seven glaciers. It has a permanent cover of snow and ice.
- The park has an outstanding natural landscape made up of a range of different landform features such as volcanic peaks, active volcanic craters, crater lakes, bare lava flows, winter snowfields, hot springs, lahar ring plains, the Rangipo Desert area, glaciated valleys and moraine deposits, river valleys and waterfalls.
- Volcanic activity, glaciation and weathering are processes that have created the landforms of the park. These processes continue to change the landforms. It is a dynamic landscape.
- Two of the North Island's major rivers – the Waikato and Whanganui – have their sources in the park.

Vegetation and climate

- Plants within the park vary from low alpine herbs and areas of thick tussocks and flax, to dense beech forests on some lower slopes. It is a harsh environment for plants. Poor pumice soils and volcanic activity slow the development of trees and shrubs.
- On the higher slopes of volcanoes the surface is made up of bare volcanic stone and volcanic gravel fields.
- There is great variation in the levels of precipitation within the park. Prevailing westerly winds plus the northeast-to-southwest orientation of the mountains result in more precipitation falling on the west-facing (windward) side of the park. Precipitation totals vary from a wet 1800 mm in the west to a much drier 1100 mm to the east of the main mountains. At altitudes above 1200 m annual precipitation exceeds 3500 mm.
- Temperatures also change with altitude. On the lower slopes of the park, temperatures vary between winter lows of -10°C and summer highs of 25°C. Frosts can occur throughout the year. Above 2000 m there are permanent snowfields and ice.

ISBN: 9780170215701

World Heritage award

- Twenty-seven sites throughout the world have been awarded World Heritage dual status. The award is given to places that are judged to have such outstanding value that they are of global importance, and therefore protecting them should be the concern of all people. Tongariro National Park is one such place. The natural features of the park and its cultural and historical significance have been judged exceptional.

Visitors

- Roughly a million people visit Tongariro National Park each year. This has grown from around 90 000 in 1960, and 500 000 in 1975.
- Visitor numbers vary across the year. The two peak times are the skiing season (from July to October) and the summer vacation period (from December to February). Summer visitors now outnumber winter visitors.
- Overseas visitors make up 3% of the visitor numbers, the rest come from within New Zealand. Tongariro is New Zealand's most visited national park.
- Visitor activities include skiing, snowboarding, walking and tramping, climbing, hunting, fishing and mountain biking. To observe, photograph and experience the environment are other major reasons for visiting.
- Types of visitors fall into one of two categories: as tourists (who visit just once or twice, perhaps as part of a New Zealand-wide tour), or as recreational users (who visit more regularly such as weekend skiers during the winter).
- Within the park there are many marked walking and tramping tracks, ski and tramping club huts and lodge accommodation, commercial skifield facilities at Whakapapa and Turoa, and hotel and motor camp accommodation at Whakapapa Village.

Fig. 10.8 Ruapehu, Whakapapa Village, golf course and The Chateau hotel

ISBN: 9780170215701

The challenge of sustainability

Visitors to the park put enormous pressure on the environment. This section considers the extent to which the environment is threatened by tourism and recreational use.

VISITOR GROWTH:
Increasing numbers of visitors
+
Wider range of tourism activities

POSITIVE OUTCOMES:
a   For tourists: more people gain 'experience and awareness of the environment'
b   For recreational users: 'their sport' – satisfaction and enjoyment provided
c   Business growth related to tourism: economic and employment growth in the area

NEGATIVE OUTCOMES:
a   Cultural offence: lack of respect shown for the environment
b   Visitor dissatisfaction because of overuse and overcrowding within the park
c   Natural environment deterioration and damage

The challenge is to manage the park so that the environment and natural processes are protected, the cultural and spiritual significance of the park recognised and preserved, while also allowing for public use and enjoyment of the park and constantly improving facilities and services for the benefit of visitors.

Perception and understanding of the national park concept varies according to the motivations of park users (Figure 10.9). For some the emphasis is on use, while for others the focus is on preservation of natural features and cultural values. Visitors often comment about the importance of protecting the park's natural heritage in order for it to be enjoyed in the future.

A: Tramping in a wilderness area.

B: Crowding at a viewing point on the Tongariro Alpine Crossing.

C: Skiing at Whakapapa.

ISBN: 9780170215701

D: Visual impact and congestion around commercial ski base facilities.

E: Boardwalks to prevent landform and vegetation damage.

F: Queues at toilet facilities within the park.

G: Road access is provided to skifield areas.

Fig. 10.9 (A-G) Tramping and snow-based activities

The **Tongariro Alpine Crossing**, a tramping track, is located on Mt Tongariro. It is famous for its unique and dramatic volcanic landscapes, and is regarded as one of New Zealand's best and most popular one-day tramping experiences. The track takes between five to eight hours to complete.

Use of the track has increased rapidly over the last 15 years. It is estimated that the Tongariro Alpine Crossing now attracts around 60 000 people every year, mainly between the months of October and May.

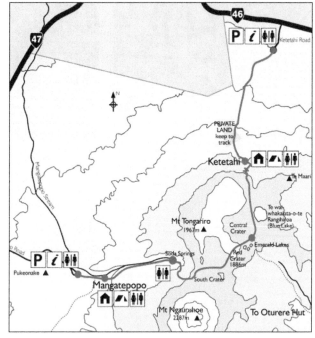

Fig. 10.10 Map of the Tongariro Alpine Crossing

ISBN: 9780170215701

An investigation was conducted to find out more about those who use the crossing. The research (primary data gathering) involved (i) discussion with track users and people with an interest in the national park, (ii) surveys of a sample of 803 people who had completed the Tongariro Alpine Crossing tramp on different days (across 28 different days), and (iii) observations made along the tramping track. The results are graphed in Figures 10.11–10.13.

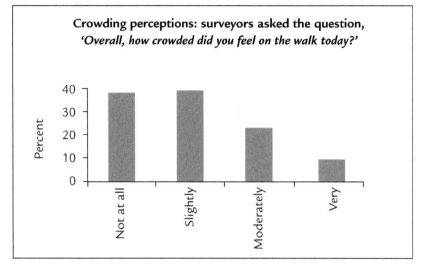

Fig. 10.11

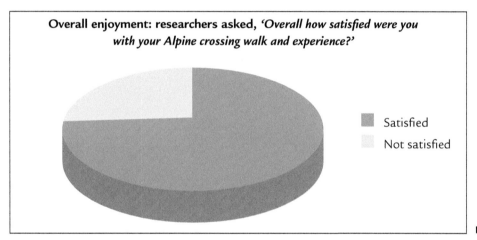

Fig. 10.12

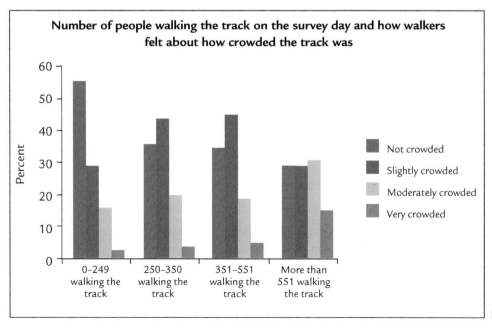

Fig. 10.13

ISBN: 9780170215701

One feature of the survey results and from the discussions with the walkers was that the feeling of being crowded and dissatisfied with the Tongariro Alpine Crossing experience was strongest among those who had completed the walk on days when there had been the largest numbers completing the walk that day (over 550 completed the crossing on the day in question).

Field research and observations suggested that trampers cause little physical damage (e.g. erosion) or vegetation loss even though the walking track is obvious and visible. In some fragile, well-used areas where walkers become concentrated, boardwalks have been built to prevent physical damage.

Cultural and social impacts of large numbers of people going on the crossing seem, at present, to be a greater threat to sustainability than impacts on the natural environment. Maori in particular have concerns about rubbish and human waste being left on the mountain. As tangata whenua, Maori also want to be consulted indepth and to have a greater say in the way the park is managed, preserved and developed in the future.

Track users often commented that if use of the track increased, then overcrowding and 'seeing too many people' would spoil the tramping experience. It would also put more pressure on the track itself and nearby facilities. Despite this, people completing the crossing are often not keen to see limitations put on the track use, as they don't want to deprive others from having the same positive experience that they enjoyed.

The Tongariro Alpine Crossing study highlights the conflicts that can arise between public use of an area and preservation of the cultural and natural environment of that area. Growth in the use of the area places extra strain on the environment, and so sustainability of the use of that environment becomes a contentious issue. On the other hand, allowing no public access denies people the right to experience these special places, and/or to enjoy their natural and cultural features.

Emerald Lakes and Blue Lakes on the Tongariro Alpine Crossing.

1 a Refer to Figures 10.6, 10.7 and the text on pages 146-147. Write paragraphs describing both natural and cultural features of the geography of Tongariro National Park and the area that surrounds it.

   b Looking at the text on pages 146-148 explain why Tongariro National Park was given a dual World Heritage site award. Include examples of specific features of the park that led to the award being given in your answer.

   c Draw an annotated sketch of Figure 10.8. Give your sketch an original title.

   d Refer to the seven photos in Figure 10.9. Judge the extent to which each illustrates impact and stress on the environment by people. Then make a large copy of the diagram on page 153 and write the corresponding letter (A-G) for the most appropriate photo in each box.

   e Justify your choices for the 'Most impact' and 'Least impact' photo choices.

ISBN: 9780170215701

Most impact and stress

Significant impact and stress

| Some impact and stress | Some impact and stress | Some impact and stress | Some impact and stress |

Least impact and stress

2 a Why do so many people visit Tongariro National Park? In your answer refer to 'attractions' and 'accessibility'. Present your answer as either a written paragraph or an annotated map.

   b Refer to the information about the Tongariro Alpine Crossing and to the track sign at the start of the walk (Figure 10.10). Create another sign that gives walkers more information about the crossing, about what lies ahead for them on this track, and other important tramping details.

3 Imagine you are attending a conference about managing Tongariro National Park in a sustainable way. Six different options (i–vi) are put forward (below). Read these six options and then complete either task a or b.

   a Choose the option for sustainable management of Tongariro National Park that you think to be the best, and justify your choice. Explain why you think it is superior to the other five options.

   b Write a newspaper editorial that provides a view about why the park needs management, and ways to manage the park that would follow best sustainability practices.

Sustainability options

   i Allow free public access to the park to carry on as at present, and continue to monitor impacts on the environment of the park on a regular basis.

   ii Allow free access to continue but provide brochures and a network of information signs within the park telling people about the natural and cultural features, their fragility and the need for respectful use of the park.

   iii Develop more facilities in the most used parts of the park (e.g. more parking areas, more toilets and washing facilities, more campsites and cabins in well-used tramping and skiing zones), but keep the rest of the park as a 'wilderness and solitude area' and as natural as possible.

   iv Restrict public access when the area becomes crowded. Set limits on the number of cars allowed in, and when car parks become full have gates that can be closed at park entrances.

   v Allow no commercial tour operators within the park, for example no bus or mini-van tours, no guiding and adventure experience operations, and no more commercial skifield developments.

   vi Keep the park as natural as possible. Restrict use of the park to activities that cannot be done elsewhere in the region. Skiing and alpine tramping, for example, would be allowed but climbing, cross-country walking, tramping, fishing and mountain biking would be prohibited.

ISBN: 9780170215701

ISBN: 9780170215701

Disappearing tropical rainforests

People have always cut down forests. Over half of the original forest cover of the earth has been cleared. The biggest clearances first took place in Europe, North America, Australia and New Zealand. Deforestation is currently happening mostly in poor tropical countries. This is where half the world's remaining forests are located (Figures 10.14, 10.15 and 10.16).

Locations of tropical rainforest

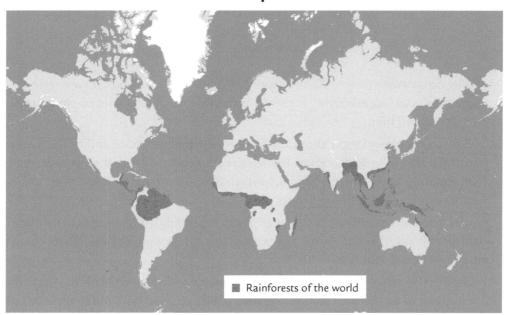

Rainforests of the world

Fig. 10.14

World land surface covered by tropical rainforest (1950, 2010)

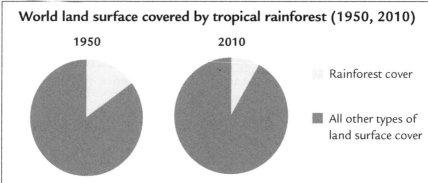

1950   2010

Rainforest cover

All other types of land surface cover

Fig. 10.15

Deforestation

The fact is that forests are seen to be worth more to people when they are felled than when left standing. Deforestation (Figure 10.17) ocurrs to make way for agricultural crops, for timber in construction, as fuel to burn, or to make travel easier.

Yet tropical rainforests have immense value: they have unique plant and animal life; they help control soil erosion and the climate; the plants hold medical cures. These things, however, generate little immediate cash and the pressure to cut remains greater than ever, in a world increasingly short of land and resources.

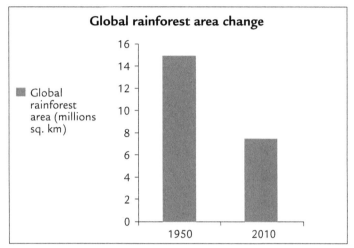

Fig. 10.16

The seven remaining rainforest areas most threatened by deforestation

1 Rivers, floodplain wetlands, mangrove forests of Myanmar (Burma), Thailand, Laos, Vietnam, Cambodia and India
2 New Caledonia
3 Southern parts of the islands of Borneo and Sumatra
4 Philippines
5 Brazil's Atlantic coast
6 Madagascar
7 Mauritius, Reunion and Seychelles islands

Table 10.3

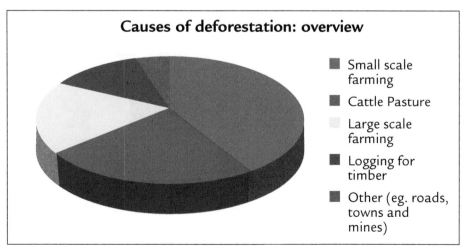

Fig. 10.17

ISBN: 9780170215701

Fig. 10.18 Causes of deforestation: ranch with Zebu cattle on land cleared from the Amazon rainforest in Brazil

Fig. 10.19 Causes of deforestation: aerial view of a large soy fields eating into the Brazilian tropical rainforest

Fig. 10.20 Causes of deforestation: timber logging in Borneo

Saving forests and sustainability

Fig. 10.21 Environmental activism: promoting recycling

Fig. 10.22 Replanting forests: a seeding nursery

Fig. 10.23 Ecotourist accommodation at Cotococha Amazon Lodge, Ecuador

Fig. 10.24 Rainforest trek with guide

ISBN: 9780170215701

ISBN: 9780170215701

ACTIVITY

1 a Refer to Figures 10.14 and Table 10.3 and, using an atlas, outline the following on a map of the world.

   i All rainforest areas, and the name of the continents in which they are located.

   ii Label the seven most endangered rainforest areas.

  b Describe the pattern of:

   i Tropical rainforests.

   ii The location of the most endangered areas.

  c Refer to Figures 10.15, 10.16 and 10.17 and the photographs on page 156 to answer the following questions.

   i What does the word 'deforestation' mean?

   ii How are rainforests changing? Write your answer in a short paragraph. Include some specific facts in your answer.

  d Why would changes like these not be sustainable? Start your answer: *'The changes that have taken place in tropical rainforests in the past 60 years cannot continue into the future because ...'*

2 a Draw an annotated sketch of one of the photos in Figures 10.18, 10.19 or 10.20. Include the words 'rainforest' plus the name of the activity that has resulted in loss of the rainforest area in your annotation.

  b There are numerous actions that people can take to help preserve tropical rainforests. Refer to Figures 10.21–10.24 and draw a star diagram that uses each photo to explain actions that will increase rainforest sustainability.

3 a Make a simple sketch of the cartoon in Figure 10.25. Label the two groups of people that are represented and suggest the parts of the world where they would live.

  b What things taking place in the cartoon would bring about changes to global climate?

  c How does the cartoon illustrate and relate to the concepts of:

   i Interdependence?

   ii Sustainability?

  d What point is being made by this cartoon? What underlying message does it have?

Fig. 10.25

ISBN: 9780170215701

Sustainable cities: Tianjin eco-city

Over half of the world's population lives in towns and cities. Sustainable urban living has become a major focus for urban planners. Achieving the goal of urban sustainability is a huge challenge for existing cities. This goal becomes more manageable when new towns and cities are being designed and built.

A sustainable city is often called an 'eco-city'. The aim of a sustainable city is to be as self supporting as possible, to make use of land and resources in the most efficient way, and to produce a minimum amount of waste and pollution. From the angle of the natural environment, creating a city with the smallest ecological footprint (natural environmental impact) possible is the aim

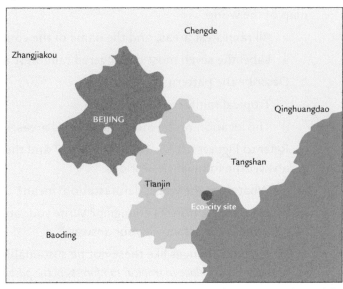

Fig. 10.26 Location and site

But sustainable cities also need to be places where the social and cultural environments meet the needs of the residents. Having a people-friendly city can be seen as equally as important as having an environmentally-friendly city in order for the urban living to be truly sustainable.

Fig. 10.27 The new bridge leading to the site area for the eco-city development

Tianjin eco-city

Tianjin is a city being developed in China as a model sustainable city. The plan is to design and build a city that is people- and environmentally-friendly – a city that works both for people and for nature. 'A city where people can live in harmony with each other, with the economy and with the environment' is the driving force behind the design. Work began on the city in 2008 and the aim is to have 350 000 residents living in the city by 2020. The eco-city is located 150 km from the capital city Beijing, and 40 km east of the existing Tianjin city centre (Figure 10.26).

Two important features of planning: 'eco-cell' and 'eco-valley'

- An eco-cell is the building block from which the eco-city will grow. In this instance it is a piece of land 400 m by 400 m. Within each cell, a range of amenities are planned such as schools, commercial zones and recreational areas as well as housing areas. With amenities located within walking distance of where people live, the need to commute to work will be reduced. The eco-cell is meant to give people a sense of community. Four or five cells will form neighbourhoods, and these in turn will become districts that make up the eco-city.

- The eco-valley is a 12 km continuous linear park that will run through the eco-city. The purpose of the eco-valley is to bring nature closer to people, and to increase awareness among the people of the need to conserve, value and protect the natural environment. This green belt also serves as the main route for pedestrians and cyclists, connecting them to major transport centres, residential areas, community facilities and commercial properties. Green corridors are planned to extend out from the eco-valley to the surrounding areas to form a network of green links within the eco-city.

A

B

ISBN: 9780170215701

Fig. 10.28
Tianjin eco-city conceptual design features

Tianjin eco-city design details
A Provide a network of public transport, cycle and walking paths. Target set for 90% of all trips within the city to use these types of transport.
B Design houses and commercial buildings to conserve energy. Include rooftop gardens in the building designs, and develop green spaces so all areas have a balance and mix of buildings and green spaces.
C Use renewable energy and clean fuel from solar and geothermal sources. Design buildings that are energy efficient, well insulated and use natural solar heat.
D Preserve existing wetlands and develop a network of both large and small green spaces within the city that are accessible from the cycle and walking paths.
E Provide a housing mix so that people with different income levels live within each community. Provide amenities and facilities within each community so they are easily accessible for all people including the elderly and mobility-impaired.
F Promote 'Reduce, reuse and recycle' among households, communities and businesses. Organic waste (e.g. paper and cardboard, food, garden and animal waste) will be used in energy generation.
G Wildlife and wetland areas will be preserved wherever possible. The existing river and the old fishing village Qingtuozi will be protected and incorporated into the city-scape.
H The area is one with low rainfall. Therefore maximum water recycling will take place within the communities and desalinated water will be produced.
I 'Clean' tertiary and service industries will be encouraged. The city will be promoted as an education and research centre for environment-related technologies.

Table 10.4

ISBN: 9780170215701

ACTIVITY

1 Use an atlas and Figures 10.26 and 10.27 to answer the following questions.

   a Describe the location of Tianjin within China.

   b Describe the site where the eco-city is being developed.

   c Match the words with their meaning in Table 10.5.

Words	Meanings
urban	Agreeable and pleasant; where things blend and work well together
goal	People who live in a particular place
ecological footprint	Towns and cities; built-up areas (opposite of rural and countryside)
eco-city	Things that surround us; natural and cultural features of our surrounds
residents	Environmentally friendly and sustainable
environment	Aims and purpose; objectives
harmony	Impact on the natural environment that people have

Table 10.5 Terms connected with sustainable urban living

   d Refer to Table 10.4 about design features of the eco-city. These features cover nine different design categories:

CATEGORIES
1. Energy
2. Buildings
3. Transport
4. Ecosystems
5. Water
6. Waste
7. Economy
8. Social Harmony
9. Heritage conservation

   i   Make full page copy of the wheel design (below).

   ii   Copy each of the nine design categories into the inner (numbered) segments.

   iii Choose the statement from Table 10.4 to match each category.

   iv   Provide a summary of this information in the outer segment. Energy has been completed as an example.

   v   Colour-code each segment and give the completed wheel diagram a title.

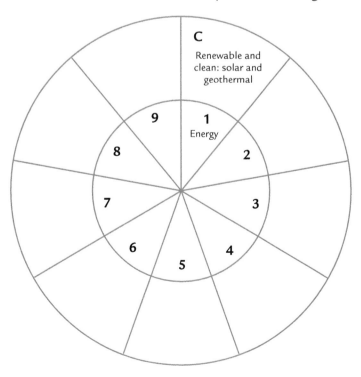

ISBN: 9780170215701

2 a Copy the 'driving force' quotation from page 158.

   b Identify and list features of the landscape shown in Figure 10.28 A, B and C in the form of an acrostic (below).

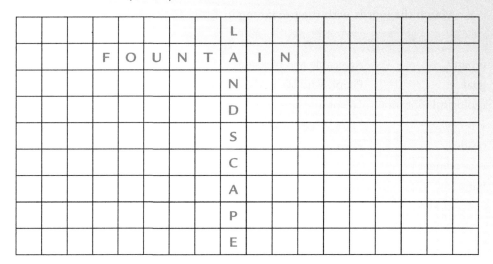

							L								
		F	O	U	N	T	A	I	N						
							N								
							D								
							S								
							C								
							A								
							P								
							E								

   c Imagine you are a radio reporter sent to Tianjin. Write a 60-second report to give listeners an impression of this city of the future based on the images in Figures 10.27 and 10.28.

   d Would you like to live in a city like this? Give reasons for your answer.

3 a Draw a sketch of Figure 10.28 C. Annotate the sketch with some design features of the city, including the eco-cell and eco-valley.

   b Explain how the urban design is intended to make Tianjin a sustainable city.

   c Use Google Earth to check on this eco-city (fly to Tianjin and from the airport head east to the coast). Find the bridge shown in Figure 10.27 and describe the scene.

11

Hidden: a story

For days, sometimes weeks, there is not a soul in sight. The sea stretches to the horizon and beyond in every direction and the fishers have only gulls for company. They count the sea as company, too, and are accustomed to its motions, sounds and mood changes. They read the sea like other people read books. Alert and concentrating all the time, the boat crew work slowly and as a team – a mistake out here could be costly, or even fatal – as there is no help nearby. The equipment isn't dangerous (if you know what you are doing), but ropes, winches and blades carry danger with them. Put them together at sea and, well ... Every fisher knows someone, perhaps a friend, who has lost a finger or arm by being careless or simply unlucky.

Despite the lack of face to face contact, radios and computers link those at sea with the shore, however it is usually at least a day or more away in sailing time. Careful attention is paid to weather forecasts as bad weather can come quickly. Reading the sea – and its twin, the clouds – are skills honed by experience. Survival depends on skills like these, which cannot be taught out of a manual. Knowing the right time to head out and when to seek shelter become second nature.

The old folk would say we have sea water in our blood. Fishing has been part of our life for generations and the whole family is involved in some way, either out at sea or onshore. We love the life – for we do not consider it work – we are carrying on the tradition. It is part of us and provides links to our past. City people cannot understand the attraction, put off by the isolation as much as the hard physical labour it demands.

ISBN: 9780170215701

Harvesting the ocean has given the family a good life. A decade ago things were especially good and catch prices kept on rising, so we invested in some new equipment and considered buying a second fishing boat. Now we are glad we didn't. To maintain our catch size we had to head further offshore and stay longer in deeper and rougher waters. And everyone who went out to sea soon began to sense that something was seriously wrong. Hidden beneath the waves.

Then came the collapse. The year when so little was caught that we had to live off our savings. Things got no better the following year. Now the family is wondering whether it is worth carrying on at all. For the first time in our lives we are thinking we might have to move from where we have fished and called home for generations. Our savings are severely run down. No one wants to buy either our boat or our gear.

Some point the finger at the large scale multinational fishing companies, roaming all the world's oceans with their modern fishing fleets and sucking fish from the sea like a vacuum cleaner sucks dirt from the carpet. These companies supply fish products and byproducts (such as fertiliser) for the international market, but unfortunately the point of no return for many fish species seems to have been reached. Quotas are either too high, set too late or else ignored entirely, and remaining stocks are so low there is little chance of recovery in the immediate future.

Throughout this century the world population has more than trebled and demand for fish has soared. Scientists report that one hundred years ago the oceans contained more than six times the fish than they do now. In 1992 cod fishing in the Atlantic Ocean off the coast of Newfoundland (Canada) collapsed because of overfishing, and more than 40,000 people lost their jobs. Cod stocks have, so far, not returned. Other fish stocks have collapsed in different parts of the world since then.

Harbour and wharf.

The abandoned factory.

We had been going back for summer vacations for years. I can remember my first trip when I was three years old and we went every year after that. We were following the family tradition, meeting up for Christmas and the New Year and camping at the beach house my grandfather had built back in the 1950s. The old town was founded on fishing. Then came new hotels and motels built for tourists, and the cafes and restaurants did a good trade.

I used to love fishing off the wharf with friends and family. We'd watch the fishing boats tie up on the opposite side of the bay, and unload their catch at the processing factory. When the wind blew our way we heard the noises and caught the smells of the factory at work. Grandad would tell us stories about the boats and about the time when the factory was first built. 'It made the town,' he said. It gave people work without them having to leave the area. 'Fishing was the town and the town was fishing,' he would say, tourism just a recent add on. Grandad liked the fishing people, he was comfortable with them. I think if he had his time again he would gladly join them on the boats.

Returning from a few years overseas, I can see the town is changed. The skyline, looking down from the road that descends into town and the harbour, is somehow different. There seem to be a lot of people in town, more locals than tourists, sitting talking or aimlessly wandering the streets, but the cafes are empty and some are boarded up. Only the pool room and pub seem to be well used. There are plenty of houses for sale, but no buyers. Most of the hotels and motels are up for sale, and one big hotel site has been completely cleared and stands empty perhaps hoping for, rather than expecting, redevelopment.

The view down at the waterfront is even more depressing – the fish processing factory stands idle, its wharf dismantled. Who knows what is hidden in the harbour waters? For years the fish processing plant pumped waste into the harbour and these days the water and beaches are polluted. Signs warn people away. Tourists quickly find new places to go to. Families like mine, who have long connections with the area, have stayed but maybe not for much longer.

ISBN: 9780170215701

1 a Give five examples of facts from the story.

   b The title of the story is 'Hidden'. The word appears twice in the story. Copy the two sentences that contain the word. Explain what 'hidden' refers to in each sentence.

2 a Make a sketch of the Past illustration. Using the Present illustration and the story to help you, identify and label 1–8 the changes that take place from Past to Present. Provide a key for the numbers.

   b List three options that the fishing family has for the future. State which option you think they should take and justify your choice.

3 Activities 2a and b focus on the concept of Change. Find examples of each of these concepts in the story:

a Pattern

b Interaction

c Process

d Globalisation

e Environment

f Sustainability

g Perspectives

h Location

i Accessibility

j Land use

ISBN: 9780170215701